Wireless Computing Primer

VERONICA A. WILLIAMS

M&T Books
A Division of MIS:Press, Inc.
A Subsidiary of Henry Holt and Company, Inc.
115 West 18th Street
New York, New York 10011

Printed in the United States of America

ISBN 1-55851-553-4

98 97 96 95 4 3 2 1

Associate Publisher: Paul Farrell
Acquiring Editor: Jono Hardjowirogo
Technical Editor: Jono Hardjowirogo
Managing Editor: Cary Sullivan
Copy Editors: Thomas Crofts
Production Editor: Anthony Washington

DEDICATION

For God commanded, saying, Honour thy father and mother: and, he that curseth father or mother, let him die the death. But ye say, Whosoever shall say to *his* father or *his* mother, *It is* a gift, by whatsoever thou mightest be profited by me.

St. Matthew 15: 4-5

Dedicated to the loving memory of my mother
Shirley Ann Felton Williams
who gave me unwavering confidence and determination
and to my father
Vernon Williams
who continues to provide me with vision and focus

Abide in me, and I in you. As the branch cannot bear fruit of itself, except it abide in the vine; no more can ye, except ye abide in me. I am the vine, ye *are* the branches. He that abideth in me,and I in him, the same bringeth forth much fruit: for without me ye can do nothing

St. John 15: 4-5

ACKNOWLEDGEMENTS

My deepest appreciation goes out to everyone who assisted me in this endeavor. To those who work for vendor companies; to those who have implemented wireless or mobile computing systems; and to those who are otherwise actively involved in this emerging new segment of the industry–I thank you. Additionally, I would like to thank my friends and family, especially Carolyn Braxton, Donna Williams and Deborah Robinson for their support. Finally I would like to thank the following experts who provided information and reviewed chapters:

Bob Euler of ARDIS

Linda R. Crane of the John Marshall Law School

Jim Hanley of Nextel

Bill Hays of Destineer

Lawrence Jones of CommQuest Technologies, Inc.

John LaRoche of Infrared Data Association

Martin Levetin of RAM Mobile Data

Ali Sharif of Melard Technologies

Richard Krebs of Motorola

Opher Lekach of Nettech Systems

Tom McCartney of Bell Atlantic Mobile

Ronald Powell of Ameritech

Dave Robins of ARDIS

Timothy Schmidt of Encore Consulting Group

Donald Warfield of SkyTel

TABLE OF CONTENTS

Section 1 The Basics of Wireless Computing

Section 2
Making it Work:
How to Plan, Design & Deliver Wireless Computing Solutions

Section 3
Case studies

FOREWORD

Wireless computing will be the next major advancement to hit the computing and communications industries since the personal computer and the cellular phone. Industry analysts predict phenomenal growth for the wireless communications and mobile computing markets. As travel increases and business operations increasingly move beyond the office, the demand for mobile data has begun to escalate. Mobile data is made possible through untethered communications and portable computers. The integration of multi-function, high performance software, portable computing devices and wireless networks makes it possible for consumers to collect, analyze and disseminate information from virtually any location, whenever they need to.

Motorola has made great strides in developing and promoting products and services which employ technologies that fuel the emerging wireless industry. Our number one objective is getting the wireless data communications industry off the ground. The solution is not PDAs versus PCs. Wireless computing represents the convergence of technologies from wireless communications and mobile computing.

New operating systems pack function and performance into highly portable communications devices, such as the Envoy® and Marco® Wireless Communicators, to deliver computing power to the mobile consumer. The Envoy® Communicator is a portable handheld device that lets mobile professionals instantly and wirelessly communicate information while on the move. Based on General Magic's MagicCap platform, the Envoy® communicator integrates two-way wireless, wireline and infrared communications into a single portable communications device. The Marco® communicator, which is based on Apple's Newton™ platform, delivers comparable functionality. These portable computing devices allow users to wirelessly access the Internet, exchange e-mail and fax -- free from phone lines.

Another member in Motorola's family of products is the Personal Messenger® Wireless Modem Card. The Personal Messenger® card allows users with PCMCIA-enabled portable computers and PDAs to take advantage of the same wireless communications capabilities.

Wireless computing requires more than hardware and wireless networks. In addition to hardware, Motorola also provides the AirMobile™ family of middleware software products which enable efficient end-to-end connectivity between clients and servers on local area networks. Motorola works with independent software vendors to ensure that application software is available to provide wireless computing solutions.

Through its family of products and vendor relationships, Motorola delivers a broad range of solutions to meet the wireless computing needs of individuals and businesses.

The wireless industry is primarily serving vertical markets today. At Motorola, in addition to serving vertical markets, we are also pursuing general business usage like wireless e-mail. Several other companies also deliver products and services which can be used as components in wireless computing systems. Motorola continues to partner with companies to identify, develop, package and deliver wireless computing solutions which meet the needs of an increasingly mobile population.

As more and more people explore the possibilities of implementing wireless computing systems, the need to understand new technologies, and how to employ them has grown. This book, Wireless Computing Primer, is a valuable resource for anyone who wants to learn the basics of wireless computing. It provides the fundamental knowledge that managers need to successfully implement wireless computing systems.

Randy Battat
Corporate Vice President and General Manager
Wireless Data Group
Motorola

Section 1:

The Basics of Wireless Computing

Wireless Computing Overview

What Is Wireless Computing?

Wireless computing is the process that enables one to receive, collect, analyze, review, and disseminate information as they travel or move about. It is mobile computing with untethered communications. The evolution of wireless computing has been fueled by consumer demand for immediate access to information. The development and delivery of wireless computing systems is made possible by the convergence of computing and communications technologies. The essence of wireless computing is the production of mobile data.

Information comes in many forms. It can be spoken, written, or visualized. Telephones have been used to transmit verbal information for many years. Fax machines, printed media, and postal services have been used to transmit written communications. Television and cinema have been used to transmit moving images. From a technology point of view, information can be placed in four categories: voice, computerized data, still images, and moving images.

Mobile data is produced when this information is placed in a format that can be easily transmitted and commonly understood by most computers. Mobile data primarily consists of numbers, letters, and still images. It can also consist of any computer-generated characters which may be used in the course of processing a mobile computing application. Current technology allows voice information to be translated into digital format for efficient communications. Voice can also be translated into text with a high degree of accuracy and efficiency. The technology for translating text into voice, however, is not yet mature enough for market acceptance. Mobile data, therefore, is considered to be the letters, numbers, pictures and voice generated text which can be put into a format that a computer can read (i.e., the ASCII character set). Pictures in the form of still images are increasingly produced by computer graphics, publishing, and other forms of imaging software.

Information stimulates thought. Thought generates ideas. The development and communication of ideas is the basis of advancement and achievement in our society. Any tool that has contributed to the creation of ideas has changed life as we know it. The telephone, radio, television, computer, and other inventions have had a lasting impact on the way we live our life. Wireless computing is a culmination of many of these technologies. It, too, will make its mark on the world as it contributes to the development and communication of ideas.

Wireless computing allows one to compute and communicate away from the home or office. It allows one to collect mobile data dynamically. It also allows users to receive mobile data from far-away locations. Wireless computing allows one to perform spreadsheet or statistical analysis, review the results, make changes and transmit reports to distant locations. Wireless computing gives one the ability to do all of these things in virtually any location.

Many people think of wireless computing as using a notebook computer, radio modem, and cellular phone to send electronic mail messages without physical connections. Existing technologies, however, allow much more functionality with a greater variety of hardware devices. Software will allow information to be directly linked to remote locations without using E-mail packages. Portable devices are pocket-sized and even somewhat larger ones can be easily worn on the body. Wireless networks other than cellular phones are able to send information remotely. There are many options for wireless computing and the choices are growing at a frantic pace.

For individuals, a growing number of home and office functions can now be performed while traveling. For companies, the process of getting information to and from their field forces can now be expedited and simplified. In the coming years, wireless computing will change the way that we live and work.

Components of a Wireless Computing System

A wireless computing system is the vehicle for the production and transmission of mobile data. The many components required to construct a system can be grouped into three categories:

1. Software
2. Hardware
3. Networks

To understand the components that are required to build a wireless computing system it is helpful to define the flow of information.

The Flow of Information

The timely and accurate receipt of information is the primary need that creates the demand for wireless computing. Initially, information is created from an idea or transaction. A customer may request that a package be sent by a delivery service; a manager may request a report from his or her staff; a trader may purchase shares of stock at a certain price. The pertinent information from these ideas or transactions may include the location and weight of the package being sent by the customer; or the elements to be contained in the report for the manager; or the name of the company, number of shares, and price of the stock being purchased by the trader. This information is communicated by writing it down, keying it into an automated system, or simply speaking it to someone or to a voice system.

Since wireless computing is based upon the movement of mobile data, this information becomes a part of the wireless computing system at the point that it is translated into a format that the system can comprehend. That is, the information must be placed into a computer file format. The most common format for storing numbers and letters is in an ASCII file. Popular file types for handling still images include GIF, TIF, and JPEG. Since there are numerous protocols and formats that the industry employs to handle data, the information may be translated into any number of files.

Putting information into a computer file format is a critical point in the flow of information. Information may be input into the computer by the person who produced it or by a third person. The customer's request may be input by an order entry operator; the manager may enter the request directly into a computer linked to the wireless computing system; and the trader's purchase may be input by a data entry clerk (often verbally communicated by a "catcher") on the floor of the stock exchange.

The initial point of entry for mobile data may take place at a keyboard, from a stylus to a pen device, from buttons to a handheld device, from a bar wand that scans data from a label, from a radio modem or other peripheral device, or from a host computer which sends data to a mobile device. Regardless of the point of entry, mobile data is created when someone provides information which is translated into a format that a mobile device can receive. The information is received via the hardware interface (e.g., keyboard, pen, or button) or via a communications port (e.g., serial, parallel) on the hardware device.

Once the information is received, it is processed by the portable hardware device. The information is either received in the form of mobile data or becomes mobile data as it is received by the portable hardware device. The software on the portable hardware device processes the mobile data and determines the next course of action. In the course of processing the information the software might re-sort it, use it in formulas to create new information, or simply put it in a different format for use in other parts of the wireless or mobile computing system. Based on instructions from the mobile consumer, the software then determines the next step. The information may remain resident in the portable hardware device or it may be sent to another location. If the information remains in the portable device, it may be transferred to another device or location at a later time or simply copied to tape or disk to provide a backup.

The next step in the flow of information is transferring mobile data from the portable hardware to the next link in the information chain. Mobile data may be sent via a physical connection to a portable printer, a desktop computer or docking station, or via a modem to a landline or wireless network. The physical connection to the hardware could be an infrared link, a parallel port, serial port, RJ11 port, or other communications port. The software on the portable hardware directs the mobile data to the designated point of transfer and places it in a format that the transfer medium can understand.

The software and hardware in the portable device communicate with the transfer medium to format and send the mobile data to its next destination in the flow of information. If the information is sent to a static location such as a printer or desktop computer, its mobility stops at this point. In many cases, however, the mobile data is sent over a network to another location in the information flow. By the time the mobile data arrives at the network it has been bundled with navigational information to allow the network to send it to its next destination.

The network may send the mobile data to another mobile device or to a host computer. Most networks are "intelligent" and, thus, have the ability to process mobile data during its transmission. The navigational information that is bundled with mobile data may be combined with information from the subscriber profile to determine what processes should take place during transmission. For example, mobile data may be duplicated and sent to multiple subscribers on the network. This is called broadcasting. The mobile data could be identified as having a low priority and saved for transmission at a later time or to an alternative destination. This can help the mobile consumer to better manage system resources and costs. To enhance the value of the mobile data, it may be combined with other information resident in the network before being sent to its next destination. Both landline (land based) and wireless networks may be intelligent. Either may be used to augment the content of mobile data along the flow of information. Regardless of the intelligence that is contained in a network, its primary role is to transport mobile data from one place to another.

After mobile data has flowed through a network, the next step is its delivery to the next point in the information flow. Mobile data can be delivered to another mobile consumer or subscriber connected to the network, or to a communications gateway on the network. If the mobile data is delivered to another subscriber, that person must have a computing device equipped with software and hardware to enable communication. This allows the "handshake," or transmission, to be completed. If the mobile data is delivered via a communications gateway, the destination point must be able to communicate with the gateway. The mobile data must be delivered through the gateway in a format that can be understood by the receiving party.

The final step in the information flow of a wireless computing system is delivery of the mobile data. Many wireless networks disassemble data from each source and then reassemble it at or near the destination point. If the destination point is a host computer, mobile data is often received from many mobile consumers. This data gains substantial value after it has been received and consolidated. Therefore, the mobile data does not reach its final stage until the results of consolidation have been tabulated and accessed. At that point, the results may produce a new set of mobile data.

In order to complete the flow of information, the components at each end must be able to communicate with each other through the network. The configuration of software and hardware at the destination point is often similar, if not identical, to the configuration at the initial point of entry. The information is received via a communicator which is attached to a communications port (e.g, serial, parallel) on the hardware device. If the connection is from a landline communications network, its destination may be a front-end processor which is channel-connected to a mainframe computer. If the connection is from a wireless network, it will be received by a radio modem connected to a computer, or a communications link which is connected to a gateway on the wireless network. In either case the mobile data travels through the communications connection and is managed by the software on the receiving computer. The software directs the processing of the mobile data including translations, if necessary, and any required transfer to the host application(s). These applications then process and consolidate the mobile data for access by those using the host computer.

Understanding the flow of information makes it easier to understand the system components which make this flow possible. Each component plays a role in ensuring that all information arrives at its destination point, without errors and in a timely manner. This is a basic step in learning about wireless and mobile computing systems. The task that a wireless or mobile computing system performs is the automation of the flow of information.

Figure 1.1 depicts the flow of information.

As you view and understand the flow of information, the components of a wireless or mobile computing system become apparent. Software, the essence of all computing systems, is used to manage the movement and processing of mobile data. Hardware provides the software and network with a platform on which to work. Networks provide the conduit for the flow of mobile data.

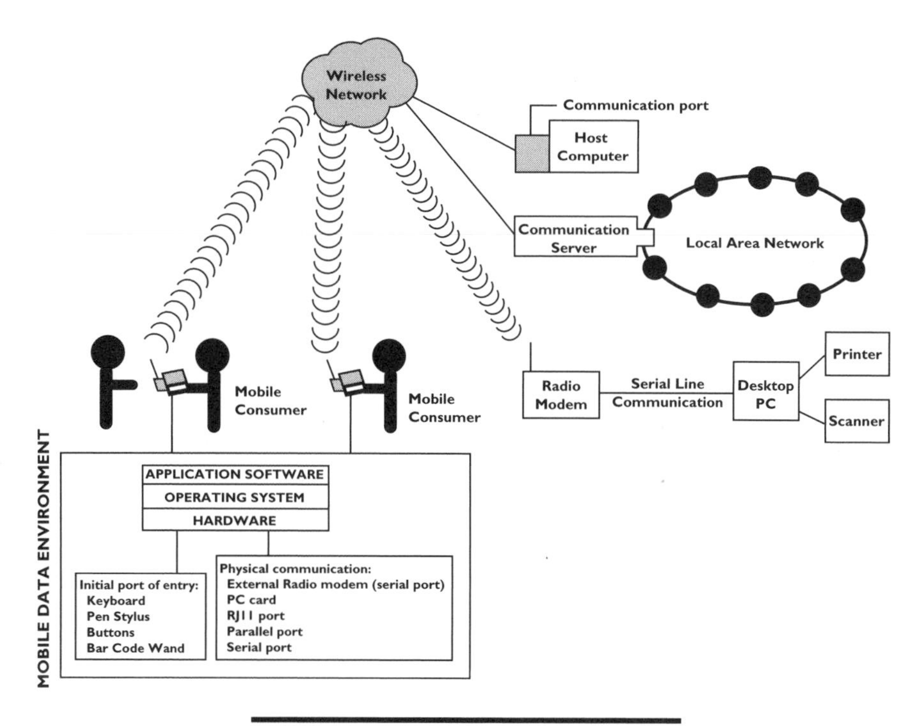

Figure 1.1 The flow of information.

Software

Software provides the intelligence that makes wireless computing systems work. Applications, systems, and communications are the primary categories for classifying software. These categories are discussed further in Chapters 3, 4, 5, and 9. Software permeates the entire wireless computing system. It resides on the portable computing device, in the network, and on the host computer. Software constructs the screens that the mobile consumer sees. Software handles the manipulation, calculation, and display of mobile data. This is also the component that directs the transmission of mobile data over the wireless network. Software manages the use and performance of all hardware used throughout the system. It is the component that drives every facet of a wireless computing system. In short, software provides the knowledge which manages the production and flow of mobile data.

Hardware

Hardware is the physical component that provides the tangible reality of wireless computing. It includes notebook computers, palmtop computers, personal digital assistants, host computers, radio modems, antennas, relay stations, and other computer and communications equipment. Hardware interfaces such as keyboards, pen styluses, and bar code scanners allow the mobile consumer to begin collecting information. Hardware is the structure that houses the software and the mobile data. It is the tool that the mobile consumer carries. Hardware provides the conduit for the flow of mobile data. It provides the receptacle that projects and captures the radio waves or electromagnetic pulses carrying information through the air. Hardware sustains the central repository for the collection and maintenance of mobile data. It is the component of a wireless computing system that most people are familiar with.

Networks

The *wireless network* is the component of a wireless computing system that is least understood. Most wireless networks are assembled with equipment (e.g., PBXs) also found in landline networks. As a matter of fact, landline networks are an integral part of many wireless computing systems. They are often part of the path to the host computer. Most wireless networks employ landline communications as part of their network architecture. It is a network's ability to receive and transmit radio waves that makes it a wireless communications network. Wireless networks can communicate short- or long-range. In other words, there are wireless wide-area networks as well as wireless local-area networks. Many different technologies are used to provide wireless communications. They include cellular, packet radio, paging, personal communications services, spread spectrum and infrared.

Several characteristics are typical of wireless networks. For the most part, these attributes are in contrast to landline networks. *Bandwidth*, or the amount of information that can be sent, is limited for the most part. Our airwaves are a limited commodity. Higher bandwidths are generally reserved for radio and television broadcasting. Wireless networks which handle mobile data are usually packet-based. This increases the efficiency and reliability of sending data over radio waves. Wireless communications *are* susceptible to interference. Weather elements and traffic from other channels can affect transmissions. Wireless communications are also terrain-sensitive. Hills, valleys, and other land formations can affect the transmission of radio waves. It is generally more expensive to send data over wireless networks; and with the exception of spread spectrum, security is sometimes an issue since radio waves can be easily intercepted. Finally, a ubiquitous wireless network does not yet exist.

These characteristics have little effect of the value of wireless computing, however. Time-critical information is not voluminous. Limited bandwidth, therefore, does not pose a major constraint. Many mobile consumers have learned from their use of cellular phones to compensate for interference and terrain-sensitivity. Productivity gains and the time-value of information, coupled with efficient systems, provide adequate cost-justification for the expense of wireless transmission. Encryption and other techniques provide a level of security which is sufficient for most applications. Finally, coverage is available where most mobile consumers need it.

Wireless networks consist of hardware and software which manages the movement of electromagnetic pulses through the air. It is the wireless network that adds the sizzle to *mobile* computing by converting it into *wireless* computing. The wireless network is the single element whose integration with portable computers signals the ultimate convergence of computing and communications. It is the addition of the wireless network that gives rise to the Computing Communications industry.

Host Computer

The *host computer* serves as a repository or resting place for mobile data. It could be a mainframe, minicomputer, or even a desktop or portable computer. The host computer may be maintained by a MIS department, service bureau, the network carrier, or the mobile consumer. For those who use only one computer for the office, on the road, and at home, the host computer and the mobile computer will be one in the same. In addition to serving as a repository, the host computer is the point of information consolidation for systems which comprise several mobile consumers. Applications used to manage and govern the enterprise are housed here. In many wireless and mobile computing systems, mobile data is not of value until it has been received and processed at the host computer.

When integrating a host computer into a wireless computing system, primary consideration should be given to the applications that will ultimately use mobile data, the communications interface, and the procedures in place to allow smooth operation of the host.

All applications that will ultimately receive or send mobile data must be identified. In addition to recognizing the specific data items which will populate the mobile data, the developer must establish how often the data items should be updated and define all calculations that need to be performed before updating data items in the host application. This information is necessary to maintain the integrity of the host application as well as the wireless or mobile computing system. The timing and synchronization of information requests can significantly impact the value of a response. Order requests, for example, may need to be delayed if they are made shortly before an update of inventory is available. Calculations made with mobile data received may also affect updates to the host application. For example, tally sheets need to be compiled from all interviewers before survey results are tabulated. The results of the entire group may form the only valid data to be transferred to the host application. The host application may need to perform certain functions before initiating transmission of mobile data. A major movement in the price of stock may require that notice be sent to brokers who are holding the stock. A thorough understanding of data items in host applications is essential to maintaining the accuracy and value of mobile data.

In order to complete the process of transferring mobile data to and from the host computer, the communications interface must be understood and integrated with the wireless system. The vast majority of minicomputers and mainframes do not have communications interfaces that readily accept mobile data. Even most local area networks and desktop computers require additional software and hardware to accept mobile data. The communications links that already exist or can be purchased for the host computer must be identified and defined. Whether it is a 3270, AS/400, DEC VAX or simply a serial link, the protocols, physical connector, and speed of the communications interface must be understood.

In order to integrate the host computer into the wireless or mobile computing system, the communications interface must be connected at the optimal location. Among other criteria, this location should allow the mobile consumers to access the host computer with the greatest ease and by the most direct route. To complete the connection, translation software and hardware adaptors may be required. This is the case when the network location does not support the protocols and physical connectors of the host computer's communications interface.

To achieve an efficient and error-free communications link, the developer must carefully select, and thoroughly test, all software and hardware components forming the connection. If the link is to a local-area network it is advisable to use a separate communications server to handle the connection. If the host computer is a desktop, the communications link can usually be provided through a serial, parallel, or infrared port using software drivers provided with the hardware or other off-the-shelf software such as Laplink. Establishing an accurate and efficient communications link is essential to the success of a wireless or mobile computing system. In most cases, this link allows mobile data to be received in the field or immediately shared with others who may benefit from it. This is a critical step in providing mobile data its time-value. The communications interface is also a critical element in enabling the end-to-end flow of information. It is therefore a major determinant of the speed of a wireless computing system.

If the host computer is part of an enterprise computing system, the developer should learn the procedures governing the operations of that system. Compliance with the operations procedures will facilitate development, maintain system integrity, and assist in successful performance of the wireless or mobile computing system. Among other things, the developer will learn how often updates take place, when backups are performed, and security processes. This information will allow the developer to plan and develop the wireless or mobile system to adapt to the rules of the host environment. Remember, when the host computer is part of an enterprise system its role is not simply to service the mobile consumer. It exists to serve the whole enterprise. Compliance with existing operations procedures, therefore, will aid in the process of integrating the wireless or mobile computing system with the host computer.

The host computer is a vital component in a wireless or mobile computing system. Few applications achieve their maximum value without it. Although many wireless networks offer peer-to-peer communications, this feature does not facilitate the enhancement of mobile data or allow the enhanced data to be immediately shared by many. In addition to adding value to wireless and mobile computing applications, the host computer often presents a major challenge in development. The total integration of the host computer is necessary in maximizing the value of a system. Differences in software and hardware may require customization, particularly for mainframe hosts. This challenge, however, has been surmounted by many who are successfully using wireless and mobile computing systems.

Benefits of Wireless Computing

Industry analysts predict phenomenal growth in wireless computing. Forecasts for the number of mobile data devices, as well as for revenues from wireless communications, show steady increase. Compound growth rates are projected to exceed 35%. This means that not only will more people be carrying portable computing devices, their use of wireless computing will also increase.

These growth rates may be considered viable given the increases in wireless network subscribers and the installed base of mobile data devices. In 1994, the paging industry enjoyed a 20% increase in subscribers, the number of customers in the United States alone reaching 19 million. By 1995, there were more than 24 million existing cellular phone customers in the United States and more than 28 thousand cellular phones were being sold every day. Although revenues were not available, companies providing services over packet data wireless networks reported a substantial increase in subscribers during 1994.

The Federal Communications Commission made history by raising an unprecedented $8.3 billion from the auction of wireless licenses from June 1994 until January 1995. This figure is remarkable considering that in the past FCC licenses were awarded based on the luck of the draw. FCC licenses awarded prior to 1995 were distributed based upon a lottery system. The last cellular lottery was held on October 19, 1994. At that time, the lottery took place in the Commisioner's Meeting Room at the FCC. It was conducted like some of the state lotteries are today—with ping-pong balls. The winners, who received licenses for radio spectrum, were those who were lucky enough to pick the right number. Some of the winners sold their licenses for substantial profits after the lottery.

What a difference ten years of competition and federal budget deficits have made. Since the last lottery, several developments have taken place. AT&T was dismantled to settle an antitrust suit initiated by MCI. The breakup of the Bell System allowed several industry segments to flourish. The long distance market opened up as new companies entered the field. These companies now hold a respectable share of the long-distance market as they control billions of dollars in service revenues. Making a concession (whose value was disputed internally) AT&T handed the undeveloped cellular business to its offspring—the Regional Bell Operating Companies (RBOCs). Although cellular phone service became available with limited coverage in 1983, its usage did not become widespread until 1992 when revenues began to take off. From 1992 to 1994, cellular usage grew from 8.9 million to 19.3 million subscribers. Revenue growth was 56% during this period.

In addition to the breakup of AT&T, several other changes took place after the last lottery which have impacted the wireless computing market. Cable companies have enjoyed record growth as the number of U.S. households that have this service grew from 17 million in 1980 to 59.1 million in 1994. In addition to subscriber and revenue growth, mergers and acquisitions have created cable companies like Tele-Communications Inc. (TCI) and Cox Communications Inc. (Cox) with billions of dollars to invest. These companies have joined the pack of those seeking a stake in the wireless communications market. A consortium which included Cox and TCI won the Personal Communication Services (PCS) license for the lucrative Northeast region which includes New York City and surrounding areas.

From software and hardware to communications and information content providers, companies throughout the information management industry are exploring or moving to stake their claim in the wireless computing market. Why? What is the draw that's attracting billions of dollars to this not-yet-established market? Simply put, it is because wireless communication enables the movement of mobile information.

Information, particularly timely information, provides power. To be certain, there are other forms of power. Money is power—but knowing when and where to invest it, or disinvest it, provides one with the power to make or break the money owner. Nuclear weapons are power—but knowing where they are located, and how to disarm them, substantially weakens the position of the owner of nuclear weapons. Presiding over a major country is power—but knowing how to influence and control the people over whom one presides will make or break that position. Knowledge is power.

Information becomes data when it is placed in a format that computers can read. Data becomes mobile when it can be transported from one location to another. Information is the critical element that fuels the power of knowledge.

Wireless computing enables the movement of mobile data. What does this mean to the consumer? The benefits of wireless computing can be grouped into two categories:

1. Productivity
2. Time-Value of Information

These benefits translate into customers who are more informed, employees who accomplish more in a given amount of time, invoices that are collected faster, or money that is spent more wisely. In short, these benefits mean a high quality of service and increased profitability for most consumers.

Productivity

Productivity has long been considered a reason for purchasing computing systems. For many years it was accepted, often without question, that computers allowed one to accomplish more in less time with fewer resources. By the mid-80s, the personal computer industry was in the midst of a revolution as computing power was placed in the hands of the masses. New products from computer chips, software, local-area networks, and peripheral devices engulfed the market. As competition intensified, new products began to emerge every three months. Confusion set in among consumers. Consumers began to step away from the frenzy and evaluate their purchases. Many found that while benefits were achieved, the productivity gains that they expected had not been realized. As a result the market has become more aware of a step which should have never been neglected: the need for defined *productivity targets* became apparent. It is important when evaluating the purchase of a wireless or mobile computing system to define and set measurable productivity targets.

There are as many measures of the productivity of wireless computing systems as there are applications and people who can benefit from them. To define the productivity gains that may be realized from a wireless computing system, it is helpful to review the *flow of information* (see Figure 1.1). To begin defining productivity measures, determine the number of resources consumed at each point along the flow of information. These resources can be time, manpower, capital, customer goodwill, or even the job satisfaction of employees. To determine the number of resources consumed, select a measurement scale or criteria. Whatever the level of ambiguity, select a measurement with the highest level of acceptance among everyone involved in the information flow. Avoid advanced formulas and techniques such as multivariate analysis and stepwise regression—adhere to the KISS principle. Keep It Simple, Stupid!. A simpler method is more likely to be universally understood and, therefore, embraced by all. Once the flow of information has been mapped out, with the designation of all points and levels of resource consumption, the groundwork is laid for the determining of the potential productivity gains.

In order to find how resources may be saved, one must first understand how they are consumed. For each point of resource consumption determine exactly what resources are used, who uses them, and how the process of consumption contributes to the ultimate objective. Areas for savings may be readily apparent. During the course of communicating information time, labor, and physical resources are consumed. Their consumption may be limited, or even eliminated, by modifying the process that uses the resource. Training or carefully assigning those who use the resources may also contribute to savings. The review of resource consumption should take place at each point along the flow of information.

To better understand how this process can identify the savings to be achieved from wireless or mobile computing, let's look at the pharmaceutical industry. Companies spend billions of dollars developing drugs which, if patented, only provide the company with exclusivity for 17 to20 years. In addition to advertising and promotion, drugs are sold as a result of doctor's prescribing for them. The industry has found that doctors are more likely to prescribe drugs that they have had experience with. Since doctors often give patients a drug sample before prescribing it to make sure they will have no adverse affects, the distribution of complimentary drug samples has become an accepted part of the sales process. Each drug sample must be tracked and verified. At the beginning of this process, the salesperson collects a sample form from a doctor. The Food and Drug Administration (FDA) requires that the physician's I.D. number, signature, and other information is collected before drug samples are distributed. Pharmaceutical companies can also use this information to track sample distribution against sales, among other things. Pharmaceutical companies spend millions of dollars to collect and process these sample forms, and the information they contain passes through many hands before it reaches those who use it. By completing the sample form on a mobile computing device, a pharmaceutical company could realize significant savings and substantially reduce errors. These savings could be realized by reducing or eliminating the resources consumed in the process of transmitting the sample information, in the form of mobile data, from its point of collection to the systems which store this data and produce reports. (A major issue in automating this process is the acceptance of signatures in digital format; however the acceptance of the digital signature standard (DSS) may diffuse this issue.)

Once the resources to be reduced, reallocated, or eliminated have been identified, the potential savings should be quantified. First determine the amount of the resources that will be saved at each point along the information flow. Remember, the resources saved may be in the form of time, labor, or material goods. The translation of information into mobile data usually speeds the process of transmitting it from one point to another. The amount of time that is saved can be tabulated. The creation of mobile data at the point of information collection often reduces the number of people involved in the processing of that information. It is no longer necessary to move paper from one location to another. Data entry operators no longer need to input information from that paper. The number of man hours saved can also be tabulated. Paper, postage, data entry equipment, and other physical resources can be saved by using mobile data. These items can be tabulated as well.

After tabulating the time, labor and physical resources that can be saved, these savings must be valued. The cost of providing each item must be determined. The cost of labor and physical resources should be readily available. The cost in time may require more effort and creativity. The cost of labor and physical resources can be multiplied by the tabulation of time to determine their savings. Since the calculation of savings may not be a simple multiplication, be thorough in the analysis to maintain operational and financial integrity. Keep in mind that most savings are continual and will be realized as long as the process of using mobile data is in place. The value of time saved can be assessed by managers or quantified using *time-value of information formulas*. (this will be discussed later on in this chapter).

Productivity gains from wireless and mobile computing are often substantial. They must be quantified and assessed, however, to determine the real value of mobile data (see Appendix 1 for an example). The process of determining the amount of savings can be simplified by quantifying the resources used at each point along the flow of information.

Time-Value of Information

We have discussed how information begets power. The power which can be gained depends on the availability of the information and how effectively one uses it. The effectiveness of using information is dependent upon many factors which are beyond the scope of this book. The availability of information is determined by: a) who receives it first, b) how many people have access to it, and c) the amount of time in which it is limited to a single person or small group. Eventually, most information becomes available to the general public. Thus, the timeliness with which information is received is a major determinant of its value. The time-value of information is a measure of how long information holds value and the rate of decline of its value. This inherently assumes that all information has some amount of value which declines over time.

In virtually every situation, pertinent information has a declining value. Immediate notification of a movement in the price of a stock will allow a broker to take action to increase or protect the value of client portfolios. Notification of changes in stock prices is of less value if received after losses have been incurred or potential gains are limited. Being able to send a field technician to a client will allow the company to earn additional revenue. If the technician is not reached before a competitor responds, that revenue is lost. Receiving late-breaking news reports could contribute to a superior bargaining position during negotiations when all players are present. Responding to the impact of news reports may require more time and effort after the participants have dispersed. The time value of information is largely determined by the means in which that information can be used and what can be achieved from its use.

Information communicated can be written, heard, or seen. The product of communications is the transmission of information. The value of information depends on its content and the effectiveness with which it is delivered. The combination of written and visual information is a powerful means of communicating an idea. Mobile data is a format for transmitting written and visual information. The value of information, however, is not limited to its content and effective delivery. The power of communications gives information a time-value which is fueled by one's ability to gain from the information received. The time-value of information is equal to the value of future benefits that can be realized by employing information, discounted by the cost of maintaining those benefits.

Time-Value of Information

The formula for the time-value of information is provided below:

$$V_t = \frac{B_F}{(1+C)^N}$$

where

V_t - Time-Value of Information

B_F - Value of Future Benefits

C - company's cost of capital

N - number of time periods

and

$$B_F = B_1 + B_2 + B_3 \ldots B_N$$

The time value of information formula can be simplified as:

$$\text{Time Value} = \frac{(\text{Value of Future Benefits})}{(1+C)^N}$$

The major component in calculating the time-value of information is the determination of future benefits. The benefits to be gained depend upon how the information is used. Since benefits should be realized as long as the system is used, the value of future benefits is the **sum** of the benefit projections for each year. Keep in mind that competition and other forces may mandate the use of wireless and mobile computing systems. In these cases, the benefits will decline as the system becomes a requirement of doing business. In the package delivery business, Federal Express and United Parcel Service have raised the bar for customer service leaving other companies to play catch-up.

Some examples of benefits to be realized from wireless or mobile computing are provided below.

- Market Share
- Goodwill
- Efficiency

Market share is a measure of a company's dominance or relative position against competitors. Businesses can gain additional customers and; thus increase their share of the market. Companies can also increase market share by increasing the availability of certain products in the right channels. Mobile data can provide the basis for actions which can result in an increase in market share.

To quantify the value of market share, identify the *Profit per Share Point.* Market share can be measured in units, such as number of customers or products sold. It can also be measured in revenue. To determine the Profit per Share Point, the revenue valuation must go further. The costs of delivering products or services must be determined so that the net profit can be calculated. When assessing costs, be sure to include all costs necessary to maintain or protect the market share.

The net profit per share point is multiplied by the increase in market share (measured in share points) for each year that the benefits are expected to be realized. The value of future benefits (B_F) from market share is the sum of these values.

Goodwill is the result of operating in a manner which makes it easier or less expensive to run the business enterprise. It can result from rendering a higher level of service which increases customer confidence. Physical assets can create goodwill when their use creates a value which is greater than the costs of liquidating the business. The effective use of wireless or mobile computing systems can increase customer goodwill. This is particularly true for package delivery or dispatch systems. Customer goodwill can be measured by the lower costs of keeping confident customers happy. The sum of these savings for each year that they are realized is one measure of the value of future benefits (B_F) from goodwill. Customer goodwill can also be measured by the revenue from referrals. Some percentage of customer satisfaction may be achieved as a direct result of using wireless or mobile computing systems. This percentage multiplied by the sum of referral revenue for each year is another measure of B_F.

Efficiency is determined by how effectively an operation is performed with a given amount of resources. Mobile data can help companies to perform more efficiently. Resources can be deployed more effectively, sales cycles can be shortened, and steps can be taken to avert or minimize negative effects on the business. For example, dispatchers could select technicians who are closest to the customer, and best equipped to respond, to send on certain service calls. Information in answer to customer questions can be provided right away and trial orders can be immediately initiated. Brokers can sell stock before a decline in price results in substantial losses. Efficiency gains from wireless or mobile computing systems are only limited by the company's need for mobile data and their ability to successfully roll out systems. To determine the value of future benefits (B_F) which result from improved efficiency, determine the dollar amount of savings. The B_F is the sum of the savings for each year that they are realized.

Benefits that result from the time-value of information will be as widespread as the applications serviced by wireless and mobile computing systems. Managers and mobile consumers should give careful thought to the gains that can be made from the receipt of more timely information. With a little creativity and objectivity, these gains can be translated into measurable benefits. This is a crucial step in quantifying the time-value of information.

Productivity and the time-value of information provide a means of grouping the vast array of benefits which can be realized from wireless or mobile computing systems. Whether systems result in a reduction of resources along the flow of information or an increase in market share, the bottom-line benefit to the company or individual is a higher quality of service or increased profitability. These benefits are not fully realized, however, until the system has been successfully rolled out and in use for a period of time. The break-even time frame will vary with each application. In some cases the advantages are immediate. This is often the case when commonly used desktop applications are duplicated on a mobile platform. In other situations the advantages may take weeks or months to materialize. Many wireless and mobile applications which are part of enterprise systems fall into this category. It often takes time for a group of people to adjust to a new routine and work in concert with each other. For situations where mobile data is of high value, the implementation of a wireless or mobile computing system usually carries a reasonable and fast return.

How to Use a Wireless Computing System

For Personal Productivity

Mobile computing and wireless communications have been used to enhance personal productivity for over a decade. Scores of individuals have used portable computers to run software packages while away from their home or office. Applications range from word processing and spreadsheet analysis to budget analysis and marketing reports. With the advent of portable modems and communications software, individuals began to use portable computers to access mainframes, local area networks and other host systems. The portable computer became a tool for the collection dissemination and processing of remote information. Electronic mail grew in popularity and many travelers began to access it from remote locations (e.g., hotel rooms).

As the popularity of portable and remote computing grew, many wanted to be free of the confines of landline communications. The traveling, individual user, or mobile consumer as I prefer to call them, has grown accustomed to the benefits of computing on the go. Many have become dependent on using the computer as a tool to perform work while away from the office. They have also become used to communication through electronic mail systems while away from the office. The mobile consumer can prepare and submit expense (and other) reports, gather critical information without having to wait for human intervention, and send requests or direction to others in distant locations. The productivity gains from mobile computing for many are more than a luxury—they have become a necessity.

Many of these road warriors also use cellular phones on a regular basis (Figure 1.2). They avoid delays and increase their communicability when public phones are not available. Available phones are often in short supply in public places such as airports, hotels, and building lobbies. A cellular phone eliminates the need to locate and wait for a phone. Cellular subscribers no longer lose time waiting for telephones. They are free to engage in other activities when they would normally have been waiting for a telephone.

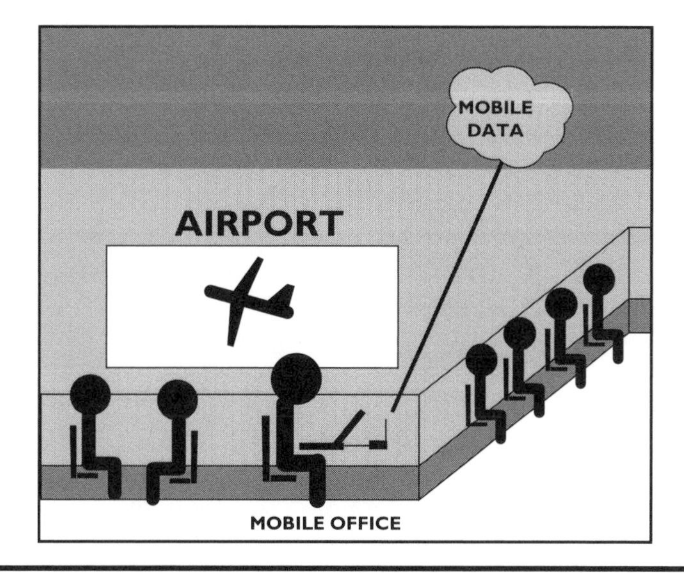

Figure 1.2 A traveling executive or salesperson uses a cellular phone and notebook computer regularly.

It was logical that early wireless computing applications for individual mobile consumers—the road warriors—would consist of portable computers and cellular phones. The introduction of cellular data modems allowed the mobile consumer to communicate with others in remote locations without having to find a landline connection. The combination of mobile computing with wireless communications offered the individual an even greater gain in productivity. The combination also allowed the consumer to save a precious commodity—time.

Figure 1.3 Business executives can better utilize their time.

Business Applications

A few major corporations have been using wireless computing systems for a few years now. You may already be familiar with some applications. Federal Express and United Parcel Service use wireless computing to track packages. General Electric and Otis Elevator have wireless dispatch applications. A growing number of people are subscribing to wireless E-mail services. Wireless computing applications that are being used, or that are now under development, cut across a broad range of industries. A few industries such as Transportation, Financial Services and Utilities have clearly recognized the time value of information. The benefits of wireless computing are not limited to these industries, however. Some of the industries that may benefit from wireless computing and the applications which yield returns are listed in Figure 1.3.

APPLICATIONS BY INDUSTRY

Transportation	- TRACKING SHIPMENTS
Utilities	- METER READING
Financial Services	- TRADING
Health Care	- PATIENT MONITORING
Newspaper Publishing	- NEWSPAPER RETURNS
Consumer Packaged Goods	- MERCHANDISING
Pharmaceutical	- SAMPLE TRACKING
Law Enforcement	- TICKETING
etc...	etc...

Figure 1.4 Wireless computing applications by industry.

Companies use wireless computing applications to send and receive critical information from employees in the field. These field employees, or mobile workers, are able to may collect information from customers which can be used to expedite billing. Information can be provided to service personnel to assist them in responding to customers faster and arriving with the needed equipment. Essentially, companies use wireless computing to arm field personnel with the ammunition necessary to perform with greater efficiency. Field personnel carry portable devices which allow them to receive and collect mobile data (see Figure 1.4).

HOW TO DELIVER MOBILE DATA

Figure 1.5 Field personnel with their mobile devices.

Rather than process paper forms continuing information that is ultimately keyed into company computers, the mobile data is sent directly to enterprise computing systems. This allows the company to expedite the flow of information while increasing the degree of accuracy. Wireless computing systems also provide an audit trail to assist in verifying the integrity of the data. Wireless computing can yield substantial benefits to businesses that rely on information from the field for their operations.

Wireless Communications Networks

Overview of Wireless Communications

History of Wireless Communications

Man has communicated by sending information through the air since the beginning of time. Prehistoric man is believed to have used smoke signals and drums to send messages to others in remote places. These methods were also used by the American Indian. Communicating over long distances was accomplished long before wire-based communications were invented. As societies have developed, the need to communicate with people in remote locations has increased. Our world has become smaller as people in different countries around the globe have worked together and shared social experiences. We rely on conventional landline based communications such as telephone, facsimile, and messenger services. But it is impossible to run wires sufficiently to provide conventional communications in *every possible location.* It stands to reason, therefore, that wireless communications was a necessary evolutionary step toward meeting the demands of a developing consumer population.

In the late 1800s, Heinrich Rudolph Hertz discovered Hertzian waves, now known as radio waves. Stimulated by his work, Guglielmo Marconi began working on a method for using radio waves to transmit signals. Marconi introduced the first wireless telegraph in 1885. England bestowed upon him the first patent ever granted for a practical system of wireless telegraphy in 1886. Just four years later, in 1900, Reginald Fessenden brought wireless voice transmission to the world. Marconi continued research and development in wireless technologies and, by 1901, transmitted radio telegraph signals across the Atlantic Ocean. Eight years later, he shared the 1909 Nobel prize in physics for the development of wireless telegraphy. By 1920 the radio broadcasting industry had emerged as intercontinental radio telegraph and radio telephone systems were in full operation and being used on a regular basis. This technology had flourished by the late 1920s and early 1930s with the transmission of pictures by wire and radio. Facsimile transmission had been born. During this time the teletypewriter produced typewritten messages with greater speed and a higher degree of accuracy than human telegraph operators could do using Morse code.

Radar developments during the Second World War made it possible to use radio at centimeter wavelengths, known as microwave radio, transmitted with beams which were only a few degrees wide. This provided point to point communication of information in larger quantities, greater speed, and higher level of reliability. By 1956 radio technology was used to send telegraphic messages and voice transmissions across the oceans from continent to continent.

Developments by Marconi, Fessenden and others had paved the way for wireless communications. By 1920 radios began to appear in homes throughout the world. Paging was introduced in 1950. Televisions began to proliferate during the 1950s. Spread spectrum technology, which was invented to provide soldiers with secure and reliable wireless communication during World War II, gained greater recognition. Cellular telephone was introduced in 1983. With the breakup of AT&T in that year, the Regional Bell Operating Companies were free to market and deliver cellular telephone service. In 1984 cellular phones were used by 91,600 people. Ten years later, in 1994, over 19 million people used cellular phones. Paging service companies began to intensify their efforts to market to the general population by the early 1990s. As the popularity of local area networking increased, so did the cost of running cables and wires. In response to the need for local area network mobility, wireless local area networks (LANs) were introduced which employed spread spectrum and in a few cases, infrared technology. By the mid 1990s, notebook computers and other peripheral equipment began to include infrared technology to allow short range transmission of mobile data.

As demands for mobility increase and technological advancement is realized, the use of wireless communications will become more pervasive. As our airwaves become more crowded, the need to control interference will become more important. Likewise, as people grow accustomed to wireless communications, the demand for worldwide wireless networks will begin to take hold. The allocation of airwaves and direction of communication technologies is controlled by government agencies and cooperative organizations. These groups develop and implement the rules and regulations for wireless communications.

Government Regulation

Anyone who has the proper equipment (transmitter, antenna, etc.) and knowledge of how to operate it can send information through the air. As you will learn later in this chapter (see *How Wireless Communications Work*), information is transmitted over different frequencies. These frequencies are determined by *the number of radio waves that are transmitted within one second.* When more than one transmission takes place over the same frequency at the same time and in the same location, interference is often the result. Interference can hamper or destroy the transmission of mobile data. As the volume of radio transmissions increased to an uncontrollable level, the interference that resulted gave rise to regulation airwaves in 1927.

The radio frequency spectrum is a finite natural resource. Although it can be used repeatedly without deterioration, the spectrum is limited. Its utilization provides a means of fast, widespread communication of information. The effective use of communications is a powerful tool in business, defense, and social welfare and it significantly impacts the way we live. The regulatory objective of many countries, therefore, is to ensure that communications is in the best interest of their citizens.

The potential gains to be realized from wireless communications are quite substantial. The Chairman of the U.S. Federal Communications Commission, Reed E. Hunt, was right on target when he stated that "the national information infrastructure is going to be a magnet for investment." This has been borne out by the magnitude of investment that has been made by companies seeking to profit from wireless communications. Over the past two years in the U.S. alone, companies have invested tens of billions of dollars in licenses, mergers, products, and services that are expected to give them a foothold in this burgeoning market. It is no wonder that the limited amount of radio spectrum has become a major concern for many. The growing demand for this limited supply is not new. In 1950 Harry Truman stated, "The most pressing communications problem at this particular time, however, is the scarcity of radio frequencies in relation to the steadily growing demand. Increasing difficulty is being experienced in meeting the demand for frequencies domestically and even greater difficulty is encountered internationally in attempting to agree upon the allocation of available frequencies among the nations of the world" [from "Allocation of the Radio Spectrum: Is the Sky the Limit?" by Sara Anne Hook, *Indiana International and Comparative Law Review*, FN10].

A few years after radio broadcasting began to flourish, the U.S. enacted the Radio Act of 1927. The Federal Radio Commission was established and given regulatory authority over all radio spectrum except those bands owned by the federal government.

Seven years later Congress passed the Communications Act of 1934 and formed the Federal Communications Commission. (The Federal Radio Commission was absorbed into the FCC). The FCC was given regulatory authority over the radio frequency spectrum and charged with granting licenses that ensured that the use of the airwaves served the "public convenience, interest, or necessity" [from "Competitive Bidding for the Airwaves: Meeting the Budget and Maintaining Policy Goals in a Wireless World," by Andrea Settanni, Catholic University of America, p.1, FN4]. The problem of how to allocate radio spectrum continued as Congress tried to come up with a fair and equitable means of dividing our airwaves for public use.

When the demand for global communications combined with the new technologies of the space program, satellite technology was born. Satellites gave us the ability to forecast the weather more accurately, improve national defense, and communicate worldwide instantaneously. Satellites were first used by the government to provide these services. Private interest in satellite service grew as companies realized the profits to be made by it. Congress passed the Communications Satellite Act of 1962, which gave FCC authority to assign frequencies for commercial satellites.

Over the years private radio systems evolved. As the owners and regulators of these systems sought to improve efficiency by sharing the operations cost among a greater number of subscribers, these systems became revenue- and profit-generating entities. The distinction between common and private radio systems began to disappear as legislative and regulatory acts went into effect. Widespread litigation arose against private radio systems. By 1982, Congress put an end to this litigation by enacting the original Section 332 of the Communications Act of 1934. Among other things, this step prevented state regulation of private mobile services and restricted common carriers from providing dispatch service.

Another communications milestone took place in 1982 when Judge Harold Greene ordered a modification of AT&T's settlement with the Department of Justice. The modification was to a 1956 Consent Decree that was granted in response to a 1949 antitrust case against AT&T. That case restricted AT&T to delivering regulated common carrier communications services. The Modified Final Judgement, which became known as the MFJ, was a response to perceived practices of AT&T that were in violation of the Sherman Antitrust Act of 1914.

Although AT&T was ordered to divest the former Bell Operating Companies, it was allowed to move into the burgeoning computer industry and keep the high profit long distance business. The Bell Operating Companies were grouped into seven regional holding companies known as the Regional Bell Operating Companies (RBOCs). These new entities were restricted from manufacturing telecommunications and customer premises equipment (CPE). The RBOCs were, however, granted the right to market and deliver cellular phone service. This provided them with an entry into the wireless communications market. In addition to cellular, the RBOCs were allowed to maintain all intraLATA (Local Access and Transport Area) business as well as provide local access to interexchange carriers (IXCs) such as Sprint and MCI. This action opened the local exchange, now managed by the RBOCs, to competing service providers. This has also been significant for the wireless communications market. The opening of the local exchange provided the primary entry into the public switched telephone network. Like Sprint and MCI, wireless communications carriers are free to access the local exchange without having to engage AT&T long-distance services.

As cellular, and other emerging wireless services, became popular and the number of wireless subscribers began to grow, concern emerged over the privacy of wireless communications. The Electronic Communications Privacy Act was passed by Congress in 1986 at the request of cellular telephone subscribers and manufacturers. This Act prohibits the reception of private communications by anyone other than the intended party. The demand for access to the radio spectrum continued to increase as companies sought to improve existing services and provide new services which employed new technologies. In August 1993, Congress passed the Omnibus Budget Reconciliation Act which included provisions to create a new class of wireless services and gave the FCC authority to grant licenses through the unprecedented process of competitive bidding (i.e. auctions). President Clinton's Budget Act paved the way for Personal Communications Services (PCS) and helped reduce the federal deficit by raising billions of dollars in license fees. The first license auctions were held in 1994. While early estimates showed that the auctions would raise $10.2 billion for the Federal government over five years, a whopping $8.3 billion was realized just from the sale of national and regional PCS licenses. The budget bill also allowed the FCC to collect fees that were estimated at $80 million annually. Clearly, the Budget Act of 1993 will continue yield even greater benefits than had been projected as auctions continue and new services are deployed.

The regulatory process continues to face challenges in its effort to keep up with technological advancements and demand for products and services which incorporate new technologies. During the summer of 1995, the U.S. House and Senate passed separate Telecommunications Reform bills. While the details of the bills are different, a major goal of each is to open all communications markets to all potential competitors. The Telecommunications Reform Bill may usher in another set of sweeping changes for the communications industry. It may modify the Modified Final Judgement (MFJ) by allowing the Regional Bell Operating Companies (RBOCs) to manufacture telecommunications equipment and offer long distance services. Cable companies and long-distance phone carries may be given approval to offer local phone services. The RBOCs may also be allowed to offer cable television and long-distance digital services. The Telecommunications Reform Bill acknowledges the convergence of the computer and communications industries by making allowances for services which are the result of combined technologies.

In short, this bill would allow long distance carriers, RBOCs, cable companies and others to provide competing services. Proponents of the legislation believe that it will result in more choices for consumers at lower prices. Opponents are concerned that inadequate competition and regulation will result in higher prices and a lower quality of service. The task which remains is consolidation of the disparate bills in a manner which satisfies the White House or the required majority necessary to override a Presidential veto. This formidable task is scheduled to be undertaken when Congress opens in September 1995.

As the pace of rules and regulations has hastened effects of its actions, the Federal Communications Commission has taken steps to increase public participation and reduce the possibility of litigation after a rule has been proclaimed. The FCC has been given the authority, under the Negotiated Rulemaking Act of

1990, to establish advisory committees to negotiate regulations which define the technical rules affecting frequency allocations. Although this process does not necessarily result in a consensus opinion which is in the interest of the general public, it does provide information may contribute to the FCC's Notice-and-Comment Rulemaking process. This process played an important role in the re-allocation and auction of frequencies for Personal Communications Services (PCS).

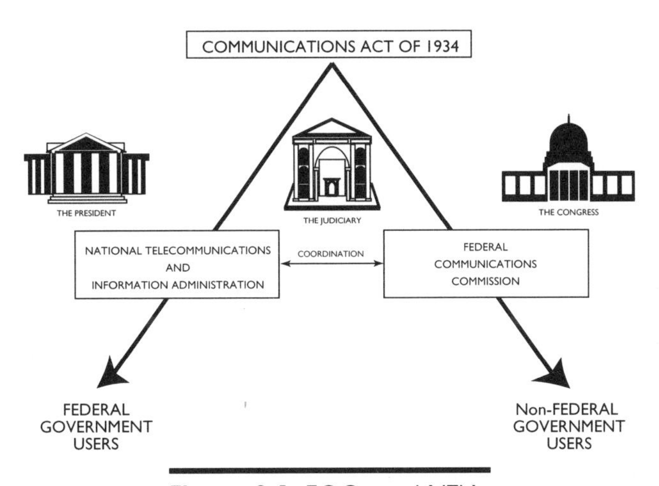

Figure 2.1 FCC and NTIA.

Two federal entities share responsibility for allocation of the radio frequency spectrum within the United States (see Figure 2.1). The National Telecommunications and Information Administration (NTIA), which is part of the Department of Commerce, represents the interests of the government and military. The Federal Communications Commission represents the interests of the remaining, larger group which includes commercial, amateur transmitters, and local government. The FCC has several responsibilities, including granting licenses for use of the radio spectrum; issuing licenses for the construction, launch and operation of satellites; and approving certain electronic devices. Its jurisdiction is the United States only. Our airwaves are worldwide, however, and frequency allocation must be coordinated with other countries to achieve global communication. To understand how this is achieved, our explanation shall begin with the organization that governs the worldwide allocation of airwaves—the International Telecommunication Union (ITU).

Originally called the International Telegraph Union, the ITU was founded in 1865 by 20 European countries who wanted to simplify the delivery of telegrams. Telephone regulation became an additional goal in 1885 and radio communication was added in 1906. The International Telegraph Union was renamed the International Telecommunication Union in 1932 and became a specialized agency of the United Nations in 1947. The chief responsibility of the ITU is to prevent radio interference between its member nations through careful allocation of the radio frequency spectrum. Although the ITU has no way to prevent unauthorized use of frequencies, it can deny legal protection to any country who causes prohibited interference to other radio services. To facilitate this process the organization requires the registration of all frequency assignments by each member country. The ITU wields widespread influence and fosters international cooperation with more than 165 member countries.

The ITU grants authority to the World Administrative Radio Conferences (WARCs) to allocate use of the radio spectrum worldwide and make changes to international radio regulations. There are two types of WARCs—general and specialized. The first general WARC was held in 1903. A historic development took place at the 1927 WARC: the radio spectrum was divided up for the first time paving the way for development of the Table of Frequency Allocations. Agreement on this table as a blueprint for domestic radio spectrum assignments was made at the 1932 WARC. The last major frequency allocation revision took place at the 1992 WARC. This was in response to technological advances that demanded spectrum already allocated. This WARC was the first major step in paving the way for a wireless communications services that employ new and emerging technologies.

The next major step in the delivery of a new class of wireless services in the United States took place when the Budget Act of 1993 was passed. In addition to the aforementioned auctions, the budget act authorized the reallocation of 200 MHz of the U.S. frequency spectrum for new services. Amateur radio operators had already lost over 100 MHz of spectrum as a result of reallocations by the FCC. They have not been affected by the last set of reallocations. To make way for the reallocation authorized by the budget act, microwave systems which operated over certain high band frequencies had to be moved. The reassignment of spectrum is a massive undertaking which costs in the millions of dollars. This is clearly acceptable, however, given the magnitude of the benefits to be expected from the new wireless communications services.

Once frequencies have been allocated, licenses must be granted to individuals or business entities who seek to provide wireless communications services. In 1991, the FCC established a Pioneer's Preference license to "encourage present and future innovators to submit proposals; to decrease regulatory uncertainty for the innovator; and to encourage investors to provide financial support"["Regulation and Licensing of Low-Earth-Orbit Satellites," by Ted Stevens, Santa Clara University School of Law]. To obtain a pioneer's preference license the applicant must present a technical innovation that will "lead to the establishment of a new service or substantially enhance an existing service" ibid]. The number of pioneer preference licenses that have been granted is small, however, and auctions have become the primary means for obtaining a radio frequency spectrum license. Although the auctions have become a popular means for granting licenses, they are predated by hearings and lotteries.

Comparative hearings began in 1945 when the Supreme Court ruled, in Ashbacker RadioCorporation *vs.* the FCC, that all mutually exclusive license applications are entitled to comparative consideration. This was in response to the FCC granting a license to a mutually exclusive competitor without a hearing while mandating a hearing for Ashbacker. The FCC used comparative hearings until 1982 to select among applicants for spectrum licenses. The applicant deemed by an administrative law judge to be most able to serve the public interest, convenience, and necessity was granted the license. Because of the high administrative burden placed on the FCC, coupled with the lengthy judicial process, comparative hearings were deemed to be too time consuming and expensive. In addition, the FCC was never able to define a clear set of criteria for selection.

In 1982 Congress gave the FCC authority to grant licenses using random selection in the form of lotteries. The intention was to speed up the licensing process and make an impartial selection among applicants. Not all entrants were equally situated to provide services, however. Some speculators made huge windfalls to the chagrin of the federal government. A typical lottery ticket cost in the hundreds of dollars. One license was sold 190 days after it was issued for $62.3 million ["Competitive Bidding for the Airwaves: Meeting the Budget and Maintaining Policy Goals in a Wireless World," by Andrea Settanni, Catholic University of America]. Three years later, however, the Omnibus Budget Reconciliation Act of 1993 paved the way for competitive bidding, or auctions.

Auctions ushered in a revolutionary way of awarding licenses for radio frequency spectrum. Many see several advantages in auctions that serve the goals of the FCC. The high cost of licenses is expected to ensure the efficient use of spectrum. Deadlines were imposed to ensure that the licensing process and delivery of new services are expedited. The auctions will also generate revenue for the government. Bidders must show their intent and financial ability to qualify with complete applications and up-front payments to bid. The auction process requires that applications be mutually exclusive, that they are for the first license or construction permit, and that they be used to provide a service for money. In striving to serve the needs of all groups, the auction plan reserves blocks of spectrum for women, small businesses, and minorities. It also provides extended payment plans and tax benefits. This goal may not be achieved, however, as these provisions may not be enough to enable all bidders to compete against well-financed corporate groups. Furthermore, auctions may not provide the opportunity for innovators with no money to deliver creative products and services. Nevertheless, auctions appear to be delivering substantial advantages over comparative hearings and lotteries.

From the allocation of spectrum to the provisioning of licenses for the operation of wireless communications, international and domestic regulation has forged ahead to foster an environment for new and innovative services. Members of several organizations have worked feverishly to facilitate the development of the new technologies that enable worldwide communications. Many of these technologies enhance the transmission of mobile data and, therefore, contribute substantially to the market for wireless computing.

How Wireless Communications Work

Information can be carried through the air via sound waves, microwaves, light waves, or radio waves. Sound waves are used to send transmissions from radio stations and other audio signals. Microwaves and light waves may be used to transmit mobile data by some networks and products that perform data conversion. Radio waves are the most popular for sending mobile data through the air.

Wireless communication networks are usually based upon the use of radio frequency (RF) technology. Radio transmissions use electromagnetic waves which are created by alternating electric currents flowing through an antenna. The electronic waves are measured in cycles. Cycles combine to form a frequency. The frequency is the number of cycles generated within one second. Wireless communications travel at different frequencies. The frequency determines how far information can travel, how much power is required to transmit it, and the functions available to enhance transmission. In many cases, frequencies are divided into channels. A channel provides a dedicated path within a given frequency for the transmission of radio waves. This is of particular benefit when sending large volumes of information over higher frequencies. Electronic waves can produce many frequencies to support wireless communications. The entire range of frequencies that can be produced is known as the *radio spectrum*.

A cycle begins with the flow of a single wave. Groups of waves combine to send data to receivers in a wireless network. Receivers identify or locate the waves when they travel within range of the receiving device. The receiver must be ready to accept the radio wave transmission. To be ready, the receiver must be tuned to recognize the frequency and characteristics of the given radio wave. When we use radios, we move the channel selector so that it may recognize the station we want to hear. Radio stations are identified on the radio selection panel by the frequency that they transmit over (e.g., 98.6). Receivers in wireless networks are tuned in a similar fashion. The receivers are also configured to recognize the modulation technique and other characteristics of radio waves intended for their reception.

Like receivers in landline networks which continually search, or poll, for information, receivers in most wireless networks accept radio wave transmissions as soon as they are in range of the receiver. A radio wave must be functional to be of use to the receiver. In order for a radio wave to maintain functionality, it must possess a minimal amount of energy, or radiation, to allow the receiver to reconstruct it after it arrives within range. Once the waves and the data that they carry are detected and received, they are then forwarded to their points of destination.

The electric current is the essence of wireless communication. To better understand how these currents form the "air waves" that carry mobile data, let's look at how waves are created in water. In its dormant state, the ocean is calm and water lies still. Forces exist that create currents which combine to move the water. As the velocity and power of the currents increase, their momentum causes the water to flow faster, resulting in waves. Small waves create ripples in the water and larger waves create visible peaks. It is common to see waves carry boats, swimmers or even bottles across the water. The waves will continue to move through the water until the force of the currents subsides or the currents fall out of reach. When this happens, the ripples and peaks created by the waves will decline until the water again returns to a calm and still state.

The same physical movements, or flows, take place when waves are created in the air. Electric current provides the force to create the flow of a radio wave. A wave begins without a flow of electricity. As the current is increased to create a flow of electricity, the wave takes form. The current continues to increase in the positive direction until it reaches a maximum flow. The current is then reversed to flow in the negative direction until the maximum, or bottom of the wave, is reached. The wave then returns to a still state, with no electrical flow. The type of electrical flow which produces this wave is called an alternating current. See Figure 2.2. The strength of the current, or amplitude, is measured by the maximum height or range of the wave it creates. Electromagnetic waves also have a pattern, which is roughly related to the strength of the radiation at any point during the waves' cycle.

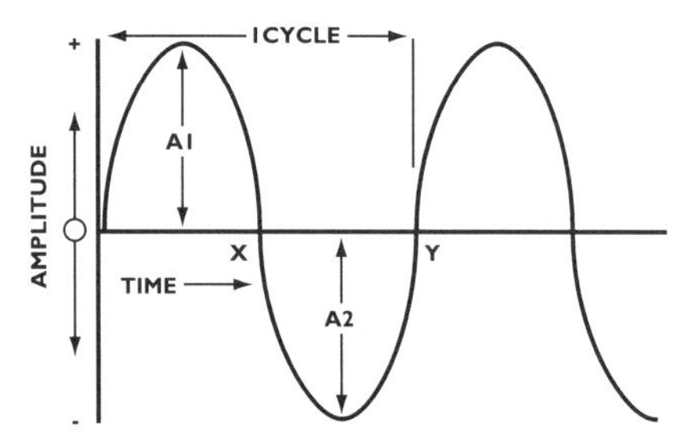

Figure 2.2 Alternating current: A1–positive direction amplitude, A2–negative direction amplitude.

A cycle begins immediately prior to the flow of current and ends when the flow has stopped. See Figure 2.3.

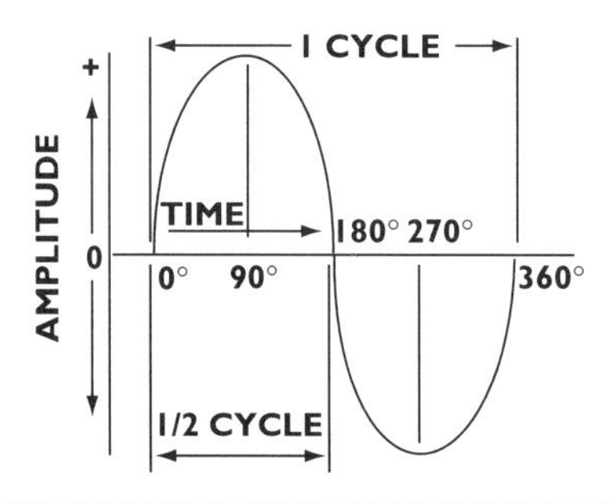

Figure 2.3 A typical alternating current cycle.

The unit of measure for each cycle is called a hertz (Hz). This measure is named after the radio pioneer Heinrich Hertz. As we mentioned earlier, the number of cycles that occur within one second determines the frequency. If 20,000 cycles take place per second, the radio waves are traveling in the 20 kilohertz frequency. Figure 2.4 shows the difference in cycles traveling in different frequency ranges.

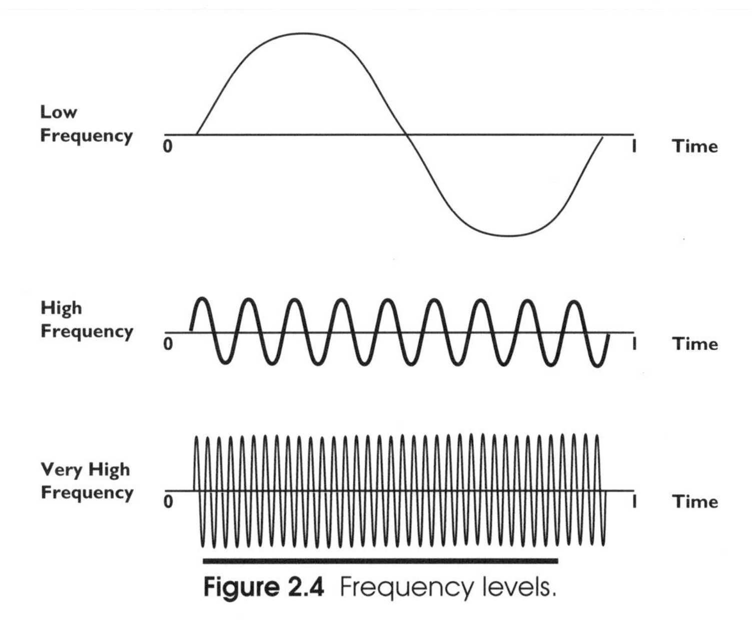

Figure 2.4 Frequency levels.

Note that the higher frequencies require a stronger current, or more power. Radio modems that communicate with wireless networks that operate at higher frequencies, therefore, require more power..

Since radio frequencies operate with a large number of cycles, they are often measured in units larger than the hertz. Frequencies can be measured in groups of one thousand cycles per second, called kilohertz (kHz); one million cycles per second, called megahertz (MHz); one billion cycles per second, or gigahertz (GHz); or one trillion cycles per second, or terahertz (THz). Waves traveling in the 20 to 20,000 Hz range are called audio frequencies. Radio frequencies (RF) are those above 20,000 Hz. The FCC allocates frequency in the 6 kHz to 300 GHz range of the electromagnetic spectrum. The U.S. Frequency Allocation Chart is shown in Figure 2.5.

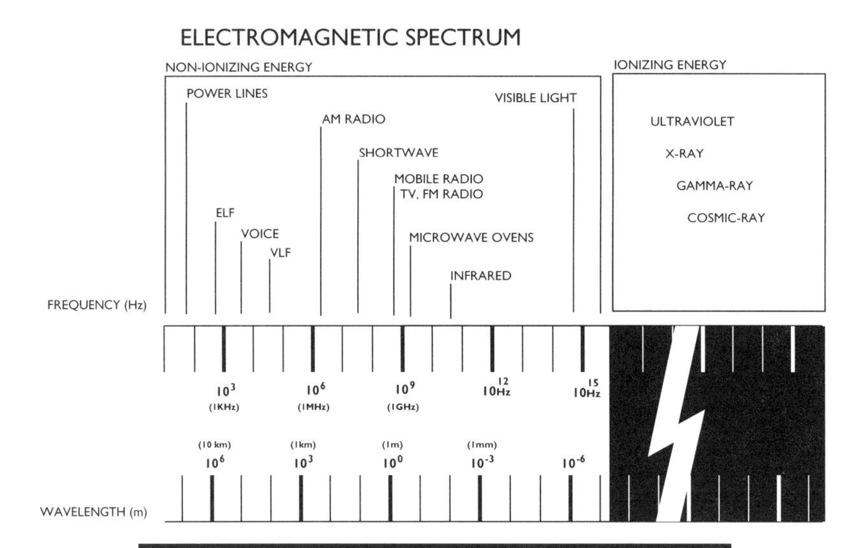

Figure 2.5 The U.S. Frequency Allocation Chart.

Radio waves begin their voyage when they are propelled into the air. This propulsion creates radiation. Radiation is defined as the propagation waves or particles, such as light, sound, radiant heat, or particles, emitted by radioactivity. Cellular phones and other wireless communications devices emit low levels of radio frequency energy, which produce radioactive waves, during their use. While concern has arisen over the safety of wireless communications due to the emission of radioactive waves, no conclusive medical evidence has been found to justify this concern. High levels of radioactive waves are known to cause biological damage. Major nuclear disasters such as Three Mile Island, Pennsylvania, and Chernobyl, the Ukraine, have proven that. Victims of these disasters were exposed to huge amounts of radiation.

Radiation sickness is defined as an illness induced by ionizing radiation. Ionizing radiation takes place when electrons are added or removed by heat, electrical discharge or other phenomena. At microwave and lower frequencies the radiation vibrates molecules but it is not enough to strip electrons off atoms. The primary effect at these frequencies is heat which is why microwave ovens work. Waves transmitted at these frequencies result in non-ionizing radiation. Xrays, which do produce minimal ionizing radiation, are part of the radio spectrum but they are very far away on the other side of light.

The radio spectrum ranges from radio waves, to microwaves, to lower infrared, to visible spectrum, to ultra violet rays, to xrays and then to gamma rays. The physics change as you move up in the spectrum so that emissions at the upper range are described as particles rather than waves. The main distinction between ionizing and non-ionizing radiations is the level of frequency. The frequencies over which wireless communications take place are not high enough to strip electrons off and produce ionizing radiation. Most physicists, engineers and biologists agree that there are not observable long term effects from radio transmissions but research continues.

The uncertainty of the health effects of wireless communicaiton devices continues to loom, however, Many engineers recommend that mobile devices be limited to 0.6 watts and that high gain antenna not be used. According to the U.S. Food and Drug Administration (FDA), there is limited evidence which suggests that low levels of radiation might cause adverse health affects as well. The amount of time that a person spends near radioactive waves, and one's proximity to them, have a major effect on the degree of exposure. The FDA suggests, therefore, that you:

1. Limit use of cellular phones to short conversations or emergency situations. Use landline phones for lengthy conversations.
2. Place the antenna as far away from the mobile consumer as possible. In vehicles, mount the antenna on ant exterior surface, far away from passengers.

Radio waves travel at the same speed as light—about 300,000,000 meters per second or 186,000 miles per second in space. The distance that a wave travels before the start of the next cycle is called the wavelength. The longer the duration of the cycle, the greater the distance that the wave will travel. In other words, the lower the frequency the longer the wavelength.

With the proper equipment we can produce electric currents to create "controlled" waves which can be used to carry information. Electric equipment can be used to create a flow of current that operates within a given frequency. That is, the current creates the number of waves within a one second period of time to generate a given frequency. The current travels through circuits from the motherboard of the computing device to the radio modem and then through the antenna. Mobile data is attached to these waves before being sent through the antenna. This takes place on the motherboard or in the radio modem. It is the essence of the wave until it reaches its identified destination. Mobile data becomes part of the electric current that generates the wave. This is accomplished with the generation of electric on and off pulses which form computer characters (e.g., ASCII set). ASCII is the native code for most computer (Figure 2.6). The FCC has established rules that provide for controlling printers and other peripherals using the ASCII character set.

The ASCII Coded Character Set

<table>
<tr><td></td><td></td><td></td><td></td><td></td><td>6</td><td>0</td><td>0</td><td>0</td><td>0</td><td>1</td><td>1</td><td>1</td><td>1</td></tr>
<tr><td>Bit</td><td></td><td></td><td></td><td></td><td>5</td><td>0</td><td>0</td><td>1</td><td>1</td><td>0</td><td>0</td><td>1</td><td>1</td></tr>
<tr><td>Number</td><td></td><td></td><td></td><td></td><td>4</td><td>0</td><td>1</td><td>0</td><td>1</td><td>0</td><td>1</td><td>0</td><td>1</td></tr>
<tr><td></td><td></td><td></td><td></td><td>Hex</td><td>1st</td><td>0</td><td>1</td><td>2</td><td>3</td><td>4</td><td>5</td><td>6</td><td>7</td></tr>
<tr><td>3</td><td>2</td><td>1</td><td>0</td><td>2nd</td><td></td><td></td><td></td><td></td><td></td><td></td><td></td><td></td><td></td></tr>
<tr><td>0</td><td>0</td><td>0</td><td>0</td><td>0</td><td></td><td>NUL</td><td>DLE</td><td>SP</td><td>0</td><td>@</td><td>P</td><td>`</td><td>p</td></tr>
<tr><td>0</td><td>0</td><td>0</td><td>1</td><td>1</td><td></td><td>SOH</td><td>DC1</td><td>!</td><td>1</td><td>A</td><td>Q</td><td>a</td><td>q</td></tr>
<tr><td>0</td><td>0</td><td>1</td><td>0</td><td>2</td><td></td><td>STX</td><td>DC2</td><td>"</td><td>2</td><td>B</td><td>R</td><td>b</td><td>r</td></tr>
<tr><td>0</td><td>0</td><td>1</td><td>1</td><td>3</td><td></td><td>ETX</td><td>DC3</td><td>#</td><td>3</td><td>C</td><td>S</td><td>c</td><td>s</td></tr>
<tr><td>0</td><td>1</td><td>0</td><td>0</td><td>4</td><td></td><td>EOT</td><td>DC4</td><td>$</td><td>4</td><td>D</td><td>T</td><td>d</td><td>t</td></tr>
<tr><td>0</td><td>1</td><td>0</td><td>1</td><td>5</td><td></td><td>ENQ</td><td>NAK</td><td>%</td><td>5</td><td>E</td><td>U</td><td>e</td><td>u</td></tr>
<tr><td>0</td><td>1</td><td>1</td><td>0</td><td>6</td><td></td><td>ACK</td><td>SYN</td><td>&</td><td>6</td><td>F</td><td>V</td><td>f</td><td>v</td></tr>
<tr><td>0</td><td>1</td><td>1</td><td>1</td><td>7</td><td></td><td>BEL</td><td>ETB</td><td>'</td><td>7</td><td>G</td><td>W</td><td>g</td><td>w</td></tr>
<tr><td>1</td><td>0</td><td>0</td><td>0</td><td>8</td><td></td><td>BS</td><td>CAN</td><td>(</td><td>8</td><td>H</td><td>X</td><td>h</td><td>x</td></tr>
<tr><td>1</td><td>0</td><td>0</td><td>1</td><td>9</td><td></td><td>HT</td><td>EM</td><td>)</td><td>9</td><td>I</td><td>Y</td><td>i</td><td>y</td></tr>
<tr><td>1</td><td>0</td><td>1</td><td>0</td><td>A</td><td></td><td>LF</td><td>SUB</td><td>*</td><td>:</td><td>J</td><td>Z</td><td>j</td><td>z</td></tr>
<tr><td>1</td><td>0</td><td>1</td><td>1</td><td>B</td><td></td><td>VT</td><td>ESC</td><td>+</td><td>;</td><td>K</td><td>[</td><td>k</td><td>{</td></tr>
<tr><td>1</td><td>1</td><td>0</td><td>0</td><td>C</td><td></td><td>FF</td><td>FS</td><td>,</td><td><</td><td>L</td><td>\</td><td>l</td><td>|</td></tr>
<tr><td>1</td><td>1</td><td>0</td><td>1</td><td>D</td><td></td><td>CR</td><td>GS</td><td>–</td><td>=</td><td>M</td><td>]</td><td>m</td><td>}</td></tr>
<tr><td>1</td><td>1</td><td>1</td><td>0</td><td>E</td><td></td><td>SO</td><td>RS</td><td>.</td><td>></td><td>N</td><td>^</td><td>n</td><td>~</td></tr>
<tr><td>1</td><td>1</td><td>1</td><td>1</td><td>F</td><td></td><td>SI</td><td>US</td><td>/</td><td>?</td><td>O</td><td>_</td><td>o</td><td>DEL</td></tr>
</table>

ACK	=	acknowledge	FF	=	form feed
BEL	=	bell	FS	=	file separator
BS	=	backspace	GS	=	group separator
CAN	=	cancel	HT	=	horizontal tab
CR	=	carriage return	LF	=	line feed
DC1	=	device control 1	NAK	=	negative acknowledge
DC2	=	device control 2	NUL	=	null
DC3	=	device control 3	RS	=	record separator
DC4	=	device control 4	SI	=	shift in
DEL	=	(delete)	SO	=	shift out
DLE	=	data link escape	SOH	=	start of heading
ENQ	=	enquiry	SP	=	space
EM	=	end of medium	STX	=	start of text
EOT	=	end of transmission	SUB	=	substitute
ESC	=	escape	SYN	=	synchronous idle
ETB	=	end of block	US	=	unit separator
ETX	=	end of text	VT	=	vertical tab

Notes

1. "1" = "0" = space.
2. Bit 6 is the most-significant bit (MSB).
   bit 0 the least-signifcant bit (LSB).

Figure 2.6 The ASCII Coded Character Set.

Each computer character is a unique combination of electric on and off pulses. These pulses combine to form an electric current. The current is modulated by the radio to form a wave which travels at a given frequency. Modulation, along with network control information, gives the wave a unique identity as it travels over a given frequency. This allows the receiving radio to recognize it as one of its own waves or signals. The radio modem reduces the density of the electric current by gradually tapering it until it reaches a slender point. This process is called *attenuation*. This allows the antenna to take control of the current.

The antenna projects the waves into the air. Antennas are resonant structures. The electrical signals produced by the current vibrate up and down the antenna in a compatible way. An antenna resonates at an electrical frequency in the same way that a tuning fork resonates. A tuning fork launches an audio wave into the air. An antenna launches, or propels, a radio wave into the air. When electrical energy moves up and down an antenna fast enough it causes electromagnetic fields to start pulsing into the air. The resonance translates it from electrical power into radio power. Antennas have a reciprocical nature. In the same manner that it takes electrical signals and convert them into radio waves, the antenna also take radio waves and covert them into electrical signals. The transceiver works in conjunction with the antenna to accomplish this. Antennas are, therefore, bi-lateral. That is, they work in both directions. The power of the device and the length and height of the antenna can be used to propel the waves to their targeted destination. The control of radio waves is lost, however, once they are released into the air. External forces such as wind, rain, or electrical interference can impair or abort the transmission. Wireless networks, however, employ technologies and practices which guard against erroneous transmissions caused by interference from external sources. For example, if all of the data in the transmission is not received properly, most wireless networks will send a request to the transmitter to resend the missing or damaged piece of information. Some networks will dynamically alter the path over which mobile data is sent to ensure that it arrives without error.

The primary function of a network is to receive information from one point and send it to another. This process may take place several times during the course of sending information. The directing of information from one point to another is called *switching*. There are two common methods for directing information in networks—*circuit switching* and packet switching. In circuit switching, one circuit is dedicated to communications between two points. A single circuit can have 64K of bandwidth. Multiple circuits can be grouped together to send larger volumes of data. The amount of bandwidth determines the space available, or size of the conduit, for sending information. Circuit switching is common in voice networks.

In packet switching networks information is divided into discreet groups, or packets. Packet switching allows information to be sent in varying quantities without monopolizing the communications path, or bandwidth, because a communications line is not dedicated between two points. The size of the packet and bandwidth that is required depends on the network technology and design. Many landline networks transmit packets within 8.64 MB of bandwidth. Due to its ability to transmit data reliably and efficiently, packet switching is common in data communications. Figure 2.7 shows a circuit and packet switch.

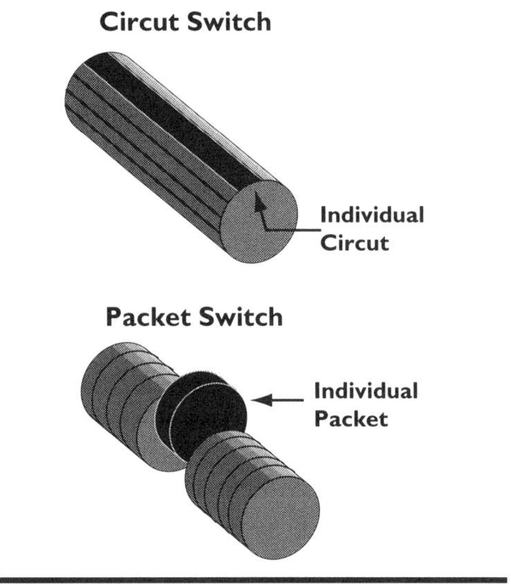

Figure 2.7 Circuit and Packet Switch.

The Deployment of Wireless Communications.

One of the major hurdles to be overcome in building a radio frequency network is gaining the zoning rights that are required to construct radio towers or erect radio antennas and their accompanying systems. A radio system is the antenna and base station that operate together to receive radio wave transmissions. Although some radio systems may be fairly compact in size (e.g., four by three feet), they must be placed in many locations, with physical communications connections, to have a broad coverage area. This means gaining approval for their placement from a vast number of people or entities. In some cases, new service providers will align with existing carriers in an effort to co-locate new radio systems with those already in place. The FCC is taking steps to control the location of antennas with a national standard and fulfillment process.

To provide wide area coverage, the radio systems are placed in locations that allow them to receive radio wave transmissions from mobile consumers within their range. The distance that a radio wave will travel depends on the frequency that it is traveling over, the technologies used to send it, atmospheric conditions, and a variety of other factors. In many networks a *cell* design is used to determine the placement of radio systems (see Cellular, Packet Radio and SMR Networks). A cell is a geographic area whose boundaries define the transmission range for a radio system. Cells may overlap to ensure that transmissions at their outer boundaries are received. The radio system which sends and receives transmissions within a cell is often called the *cell site*. These radio systems are connected via physical landline communications (i.e., wires, cables, fiber) or connected using radio frequencies. The full deployment of a wireless communications network includes physical connections between the radio systems and a network control system.

A network control system may include private branch exchange (PBX), computers, multiplexers, and other equipment that provides functionality for managing the movement and processing of mobile data. It often provides gateways for connections from external networks or systems, such as host computers, the public switched telephone network (PSTN), and other public and private networks. The network control system is the pivotal point in a wireless communications network in enabling the flow of mobile data. It provides the functions and routing necessary for the end-to-end flow of information.

The Flow of Information in Wireless Networks

Most wireless networks accept mobile data in the form of radio waves or signals. Many networks send and receive signals that contain groups of information packets. The information is created in the form of electric pulses generated by currents. The creation of these electric pulses is the beginning of the flow of information, or mobile data.

I shall define the flow of information beginning with the mobile consumer. Information becomes mobile data when it is processed by a computer or computing device. The mobile consumer begins this flow by executing a series of functions on their portable computing device. These functions are invoked using applications software to create the mobile data and communications software to send it. Communications software divides up the information and bundles it into packets small enough to be transmitted over the network. Since it originated from the portable computing device, the mobile data is contained within the packets in the form of electric pulses generated by currents deriving their power from the portable hardware. These packets of mobile data are sent to a radio modem connected to the portable computing device for transmission to the wireless network. In addition to holding the information being transmitted, these packets are structured in a defined format and contain identification information for the network. The radio modem modulates the packet by varying its amplitude, frequency, or phase. This process creates a radio wave or signal which travels over an assigned frequency. The modulation also provides information that the communications network needs to identify the signal as one of its own. A radio transmitter, within the radio modem, uses an antenna to project the signal into the air.

The radio systems (i.e., antenna and base station) in a wireless network continually search, or poll, for signals. The network recognizes a signal by the frequency that it is traveling over, its modulation, and its format and content. A signal is received by a transceiver which is located in the radio base station. The radio system (antenna and base station) is tuned to receive signals traveling over an assigned set of frequencies with one or more defined types of modulation. The radio system is also configured to receive signals using a certain access method (i.e., TDMA, CDMA). The modulation and frequency determine whether or not the signal is received. Once the signal is received, it is sent to a base station. The signal is demodulated and the information is then stripped off and sent to a network processor. The processor reads the signal. Most signals contain network and subscriber identification numbers at the beginning of the signal. If it contains the right identifying information in the right locations, it is accepted by the network.

Once the signal is received and accepted by the wireless network, it continues its journey through the network infrastructure (switches, processors, etc.) to be processed and transmitted. The network can now process the information and create another signal, which is then sent or switched to the designated recipient of the information. The signal is sent to a radio system within range of the designated recipient. It is then modulated, assigned a frequency, and propelled into the air. When the signal is received by the radio modem of the recipient, it is demodulated by the radio modem and transferred to the portable computing device.

The portable computing device now processes the mobile data into information and makes it available for use by the designated recipient, or mobile consumer. The mobile data can be used to print a report, receive a fax, or simply display information on a screen. The final step in the flow of information is in the mind of that person who uses it to produce a result or action. In the scenario described herein, that person is the mobile consumer.

Receiving Radio Waves: Modulation, Access Methods, and Antennas

Modulation is the process of altering an attribute of a radio wave so that it can be received and identified by a radio system. The altered attribute could be the *phase*, *frequency*, or *amplitude* (Figure 2.8). Phase is the amount of time that elapses between the beginning and end of a cycle. Amplitude is the height of the cycle. The frequency is the number of cycles that take place per second.

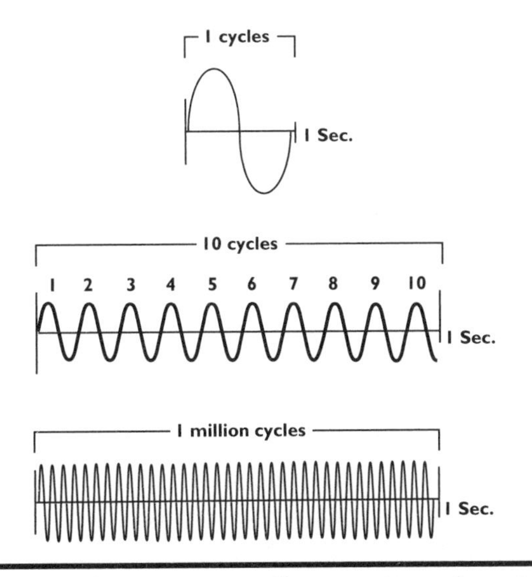

Figure 2.8 Frequency: the number of cycles

Although it is commonly referred to as a signal, a radio wave is actually a group of signals. A circuit that combines a radio wave and a modulating signal to produce a modulated radio wave is called a *modulator*. A modulator can also produce a modulating signal and send it to another circuit, which, in turn, combines it with the radio wave. The manner in which the signal is modulated identifies it to a radio system tuned to receive its specific modulation and frequency.

Sidebands are produced by the alterations produced from modulation. They exist on both sides of the frequency and are identical. This is one of the primary reasons that frequencies have space between them.

The efficiency of a modulation scheme is generally determined by how compactly it condenses the energy of the signal for a given rate of information. Condensed energy increases the strength of a signal.

The process of modulation gave rise to the term *modem*. A modem modulates a signal before transmission and demodulates a signal to receive and interpret it. Thus, the name **modem** to denote **mod**ulate and **dem**odulate. Modulation is akin to putting a network signature or identifier over a signal or radio waves. It determines which frequency the radio wave will travel over. Modulation is not the only means of differentiating radio waves between networks; however, it is usually in the second step (i.e., demodulation) that a receiver recognizes or interprets a radio wave. Demodulation removes noise and distortion as well as the modulating signal from a radio wave. This is one means of maintaining the integrity or accuracy of mobile data.

There are several types of modulation—AM, FM, and PM to name a few. Each takes a different approach to defining the attribute which differentiates a radio wave.

AM, or amplitude modulation, changes the amplitude or height of a radio wave. This method is used to transmit signals to AM radio stations.

FM, or frequency modulation, varies the frequency of a radio wave. The amplitude does not change. This method is used to transmit signals to FM radio stations.

PM, or pulse modulation, sends signals as a series of short pulses which are separated by time intervals during which no signal is transmitted. This method is restricted by the FCC to the microwave radio band.

As we discussed earlier, radio waves are commonly used to carry information in wireless communications networks. The specific amplitude is based on the power supplied by the transmitting radio's amplifier. Current modulation schemes modify the frequency, amplitude, or phase of a radio wave, or a combination of two or all three variables. Frequency Shift Keying (FSK) is considered to be one of the most robust techniques. Rather than receiving each full wave in a series separated by time, FSK receives them somewhat in parallel, overlapping into each other's time slot. This overlapping in known as intersymbol interference. One way of accommodating intersymbol interference is to send more than one fullwave per data bit. An alternative approach is to use adaptive equalizers that reverse this effect. Each method of accommodation carries a cost, either in bandwith consumption or in transmission overhead.

There are numerous modulation schemes – Frequency Shift Keyed (FSK), Phase Shift Keyed (PSK), Amplitude Shift Keyed (ASK), QAM, Multiple Shift Keyed (MSK) – most of which use full waves and synchronization (Figure 2.9).

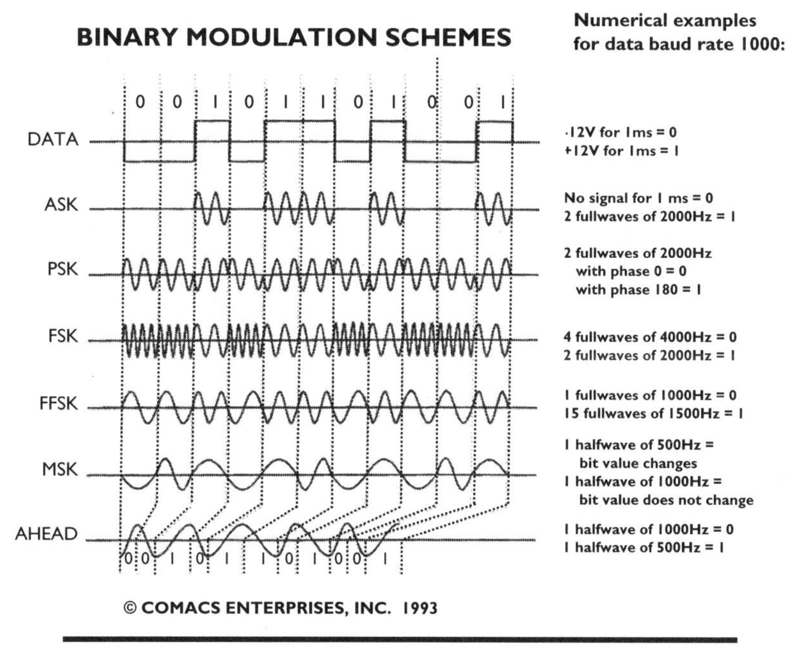

Figure 2.9 Binary Modulation Schemes versus AHEAD.

A new modulation scheme has emerged which enables more mobile data (voice and data) to be transmitted over a given frequency with moderate power consumption. The Asynchronous Halfwave Encoding and Decoding (AHEAD) technology promises asynchronous, error-free, high speed data transmission over any voiceband channel. It provides continuous-phase, frequency based digital modulation schemes, plus very high actual transmission speeds without synchronization, adaptive equalization and bandwidth occupation. In the AHEAD approach, to maximize use of the bandwith, multiple full waves or fixed time intervals are not used. AHEAD accommodates interference and depends upon only a halfwave (see Figure 2.10).

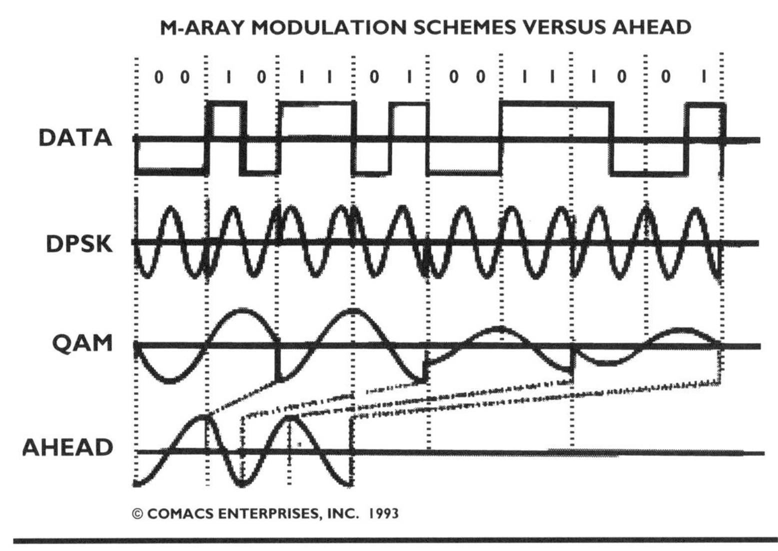

Figure 2.10 -AHEAD Modification of Carrier Signal from IDP-34 of Wireless Communications, Computing and Networking.

Since the time to transmit a fullwave is a function of its frequency, employing only a halfwave to represent a data bit cuts the transmit time in half, regardless of the frequency used. AHEAD, therefore, support a greater data rate in the same bandwidth channel consumed by the two frequencies. To compensate for intersymbol interference, AHEAD employs a patented algorithm that takes advantage of the predictive nature of analog transmissions. This modulation scheme can transmit mobile data at 10,400 bps within 2 Khz of bandwidth with no discernable effect on range. AHEAD offers a means for increasing the amount of information that can be sent over our limited airwaves – the radio spectrum. It has been certified by a Regional Bell Operating Company and a Cellular company.

There are numerous modulation schemes and many are variations of the techniques described above. The modulation scheme plays a major role in identifying radio waves to the wireless network. Once the waves are airborne, however, they must be accessed by a radio system in the network they are traveling over. As the central point for receiving signals from all mobile consumers within a given geographic area, the radio system must be able to access many signals within a short period of time so that the mobile data is received before the signal loses its strength. This is accomplished through *multiplexing*.

Multiplexing is the process of combining two or more signals or radio waves over the same channel or frequency. A channel is a portion of a frequency, or sub-frequency. Since multiplexing in a radio system allows the access and receipt of multiple signals, it is often referred to as a multiple access method.

As with modulation schemes, there are several types of access methods. Frequency Division Multiplexing, or frequency division multiple access (FDMA), is a method of transmitting radio waves over different frequencies. Each frequency or channel can be modulated in different ways. Time Division Multiplexing, or time division multiple access (TDMA), is a method of transmitting radio waves over the same channel or frequency in different time slots. Code Division Multiplexing, or code division multiple access (CDMA), uses a code sequence to transmit radio waves over different channels and frequencies. The coding sequence is used to disperse the signals and also to reassemble them. The effectiveness of the multiplexing or multiple access method plays a major role in the amount of traffic, or capacity, that a wireless network can handle. The more efficient the multiplexing method, the more signals a radio system can receive within a given time frame (Figure 2.11).

Low Mobility	High Mobility below 1.5GHz	High Mobility-2GHz
CT2/Telepoint DECT Personal Handy Phone Bellcore TA-1313 PACS	AMPS/NAMPS TACS GSM PDC PPS©-800 IS-95 CDMA IS-54 TDMA	DCS-1800 CDMA 1.25/2.5

Figure 2.11 Multiplying technologies.

The most commonly understood component in most radio systems is the antenna. Antennas have been used in everyday life since radio was introduced in the 1920s. Most of us have learned move an antenna in different directions to improve the reception of a radio or television set. We have also learned that by locating antennas close to windows we may improve reception. The antenna is the physical component in a radio system that first receives the radio wave. It is also the component from which radio waves are launched. The antenna polarization dictates the direction in which the radio wave will travel. Polarization is determined by the position of the antenna in relation to the earth. An antenna which is turned upward, at a right angle to earth, radiates a vertical radio wave. An antenna which is positioned horizontally radiates a horizontal radio wave.

The antenna's *bandwidth* is the range of frequencies that are adequately supported. This is usually determined by the range of frequencies used in the wireless network. The *gain* is the force of the propulsion which an antenna delivers. As the gain is increased, the amount of radiation is reduced. Radio waves radiate outward from a central point. As you point an antenna in the direction that you wish a radio wave to travel, you limit the distance that the wave will travel in other directions. Several types of antennas are used to transmit radio waves. They include directional, omnidirectional, high gain and smart antennas. The selection of the antenna is based upon the properties and design of the wireless network. Radio modem manufacturers collaborate with wireless network carriers to determine the best antenna for a wireless consumer. This antenna is included with, and is sometimes integrated into, the radio modem.

Wireless Computing

During its first 40 years, the computer industry increased its use of communications through landline technologies. Cables, wires, front-end processors, and other equipment were assembled to send data between mainframe computers, local-area networks, terminals, and other points at remote locations. Data communication systems emerged as the industry sought an efficient and consistent means of sending information. Data communication systems consist of three basic components: media conversion, transmission, and communications processing. These components work in concert to send information from one location to another. Data is transmitted in the form of electrical signals. Media conversion changes these signals into a format that a computer can read. *Paths* or "pipes," must exist to provide a conduit for transmitting data. *Control characters* precede, accompany, or follow the data being sent to guide its transmission, delivery, and receipt. Communications processing is the act of interpreting and following the direction of these control characters. This step translates the data into a form that the media converter can interpret. Error detection and checking also takes place during communications processing. Data communications supplies the means to move information at high speeds from one computing location to another.

In many cases, data is sent in large quantities. Technologies that have been employed for "conventional" data transfer between computers generally use session-based *communications*. These include 3270 based sessions, the public switched telephone network (PSTN), T1, and others. Session-based communications establish a bridge between the two points of data transfer. This bridge remains intact until the communications session has ended. Many users of dial-up systems know all too well how data can be lost when these bridges collapse and the connection is dropped. Subscribers to on-line systems also have learned that their costs increase with the length of the communication sessions. Nevertheless, session-based communications is a reliable, low cost means of sending data between computers.

The essence of mobile computing is the ability to process and transmit 0s and 1s, or on and off electric pulses. A "0" represents a pulse and a "1" represents no pulse. These are the core elements which comprise mobile data. The growth of mobile computing has spurred demand for wireless data communications. The vast majority of all desktop personal computers are connected to a public or private communications network. It stands to reason that mobile-computing users want network access. The conventional physical connections to communications networks, however, disable mobility. Providing wireless communications to mobile computer users was the next logical step. Thus, the convergence of mobile computing and wireless communications began to take place and wireless computing systems began to emerge.

The first wireless computing systems began to appear in the late 80s. Portable devices equipped with radio modems allowing the transmission of mobile data hit the public market. Early systems used proprietary radio networks. Companies like General Electric and Sealand used proprietary radio networks to communicate information from the field. IBM's need to communicate with their field technicians gave rise to the ARDIS wireless network. New companies were born, alliances were established, major investments were made, and research and development in wireless communications and portable computing devices increased. Executives from major corporations jumped ship into the alluring waters of those companies best positioned to take advantage of the impending growth in wireless computing.

By the mid 90s a flurry of products hit the market as companies introduced wireless local area networks, PC cards, palmtop and notebook computers, and services delivered over wireless networks. The number of wireless network subscribers increased and new wireless networks were announced. Even software companies began to jump on board as Microsoft, Lotus, Oracle, and others announced mobile and wireless versions of their software. As more people began to spend more time working away from the office, the demand for the combination of computing, communications, and mobility increased. The purchase of portable devices began to take off. Software used on mobile platforms gained in popularity. The number of subscribers to wireless networks grew as did service revenue (see Appendix II). The public was beginning to purchase products and use services which allowed them to compute and communicate while in transit. The market for wireless computing had arrived.

Cellular Networks

Distribution of Licenses

The Federal Communications Commission has specified that two cellular carriers operate in each of the top 305 Metropolitan Statistical Areas and has allocated one radio band to each—A and B. The Regional Bell Operating Companies (RBOCs) became major players in the cellular market when they split from AT&T in 1983. Under the terms of the divestiture agreement, the RBOCs—not AT&T—retained the right to provide cellular service. As the subscriber base and revenue from cellular services grew, the non-RBOC companies began to seek additional funding to increase their traffic capacities to keep up with the growing demand. In addition to major investments in these companies, mergers and acquisitions also took place. Companies like McCaw Cellular and Cellular One moved ahead of the pack as they acquired licenses to vastly expand their coverage areas. The distribution of licenses to provide cellular service decreased as fewer companies with larger service areas emerged.

As the demand for wireless communications increased, the threat of competition increased as other companies introduced or enhanced services over other radio frequencies. Additionally, the FCC allocated spectrum to provide a new class of wireless communications called Personal Communications Service (PCS). Among other things, the market began to demand simplified, single-source billing and the efficient transmission of mobile data over wireless networks. It became apparent that a collaboration among cellular carriers was necessary to meet the growing demands of an ever-increasing subscriber base. Major technical and business issues had to be resolved to allow cellular carriers to meet those demands. In 1994, the CDPD (Cellular Digital Packet Data) Forum was established by Bell Atlantic Mobile, McCaw Cellular, Ameritech, IBM, GTE, Airtouch, and NYNEX Mobile Communications. An outgrowth of the CDPD Forum has been the definition and establishment of a packet data protocol to run over the 900 MHz band as well as a plan to provide a single source of billing. The membership of the CDPD Forum now consists of several types of companies including cellular carriers, hardware manufacturers, and independent software vendors.

The distribution of licenses has decreased since they were first issued by the FCC. The size and growth of the companies who hold those licenses, coupled with the FCC requirement of two companies in the major Standard Metropolitan Statistical Areas (SMSAs), however, virtually guarantees that the distribution will remain somewhat widespread. The CDPD Forum provides a means through which cellular providers and their business partners can function as a cohesive group.

Cellular Technology

The technology used in cellular networks supports the transmission of analog voice, circuit switched data, and digital data over the 900 MHz radio band. Most cellular companies offer two technologies for the transmission of mobile data—Circuit Switched Cellular (CSC) and Cellular Digital Packet Data (CDPD). Circuit switched cellular is sometimes referred to as Advanced Mobile Phone Service (AMPS). While each technology supports transmissions over the same frequencies, the manner in which they operate is quite different.

Circuit switched cellular uses a complete path or channel to transmit voice and data. Once the channel connection has been established, it is used exclusively by the two points of communication, or subscribers. Although circuit switched cellular uses the same channels and frequencies as CDPD, the access, or modulation, method is different. (Modulation allows the radio to distinguish between signals from the two technologies.) Circuit switched cellular uses frequency modulation (FM). CSC also has priority use of available channels over CDPD except in those instances when CDPD volume is high.

Cellular digital packet data sends mobile data over the cellular frequencies in the form of digital packets. CDPD packets are typically 128 bytes in size. Each packet contains a destination address, source address, sequence number, and the data being sent. Mobile data is formatted into the CDPD protocol using communications software and CDPD modems. The access method or modulation scheme used to recognize CDPD signals is GSMR. This allows the radio to direct the CDPD signals through the network infrastructure designed for CDPD.

In order for mobile data to be recognized and received by the CDPD infrastructure, the modems must include four identifiers:

- Network Entity Identifier (NEI)
- Electronic Equipment Identifier (EEI)
- Authentication Sequence Number (ASN)
- Authentication Random Number (ARN)

In addition to allowing the CDPD infrastructure to recognize its packets, the identifiers are also used to eliminate fraud. The network entity identifier is an Internet protocol address that serves as a phone number. To recognize the specific, physical modem from which mobile data is being transmitted, the CDPD infrastructure reads the electronic equipment identifier. The remaining identifiers—ASN and ARN—are used to stop fraudulent use of the network. When the modem is first used, the ARN and ASN are processed, sent to the network to be stored in the database, and also stored in the modem. These identifiers are never transmitted again. Every time the modem is used these identifiers are processed within the modem. The ASN is incremented by a certain amount and an accompanying ARN is assigned. The incremental number for the ASN and its randomly selected ARN are sent to the network to update the database. Thus the network database remains in synch with the usage information in the modem. This makes it virtually impossible for a duplicate modem to be used.

The four modem identifiers allow the network to recognize and communicate with the mobile subscriber to whom the modem is assigned. They not only enable communication, but also eliminate fraud.

All CDPD modems use a very robust encryption scheme called RC4. This helps to protect the mobile data from interception and interpretation by a third party. In order for mobile data being transmitted via the CDPD infrastructure to be of value of an unwanted party, that party be able to reassemble the data packets and break the code of the encryption scheme. Encryption takes place at the CDPD modem and ends at the point that radio waves are demodulated in the cell site. Mobile data travels through the network infrastructure without being encrypted. To attain a high level of security using encryption, mobile data must be coded at the time that it is created and remain so until it is processed by the party designated to receive it. This degree of protection, end-to-end encryption, is beyond the scope of a network. The design of the CDPD infrastructure provides protection for mobile data at its most vulnerable point—when it is in the form of radio waves that travel through the air. The CDPD infrastructure, therefore, provides an environment which keeps mobile data fairly secure.

Since the CDPD equipment is co-located with CSC equipment, to demonstrate the flow of mobile data I will explain the design of the cellular network.

Network Design and Components

The name cellular is based on the design of the network. The geographical area throughout which companies provide cellular coverage is divided into smaller areas called cells (Figure 2.12). Each cell is served by a low-powered radio transmitter that links the mobile phone, or portable transceiver, to the cellular network infrastructure.

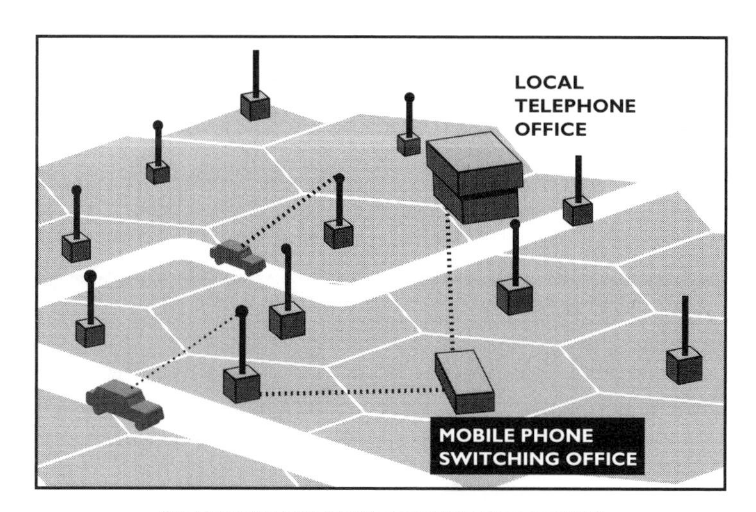

Figure 2.12 A typical cellular area.

To define the components of a cellular network, I shall explain the role that they play in the flow of information. Since a major objective of wireless computing is to equip the mobile consumer with computing and communications functionality, the flow of information shall begin with the mobile consumer. To gain access to the cellular network, the mobile consumer is equipped with a *mobile end station* (M-ES). A mobile end station is a portable device that contains a radio transceiver tuned to the cellular network. The M-ES includes a radio modem with a distinctive pigtail antenna. (It should be noted here that many modems are not equipped with a radio transceiver.) The portable computing device must be physically connected to the mobile end station. The computing device processes and sends the data to the modem via the physical connection (i.e., serial or TTY port). The modem modulates the data for transmission over the cellular network. Keep in mind that the cellular network incorporates two separate infrastructures—CSC and CDPD (Figures 2.13 and 2.14).

If the mobile data is designated to transmit over the circuit switched cellular infrastructure, the modem will use frequency modulation. If it is designated for the cellular digital packet data infrastructure, the modem will use GSMR modulation. The mobile data is sent from the modem to the radio transceiver which propels it into the air through the antenna.

Cellular transmission takes place over two frequencies which are 45 MHz apart from each other. One frequency is used to send information and the other is used to receive it. The frequencies are in the 900 MHz radio band. Although these two frequencies may change from one cell site to another, they are always 45 MHz from each other to avoid distortion and interference. Cellular communications use two channels. Each channel reside in one of the two frequencies. One channel is used to transmit from the cell site to the mobile consumer. This is called the *forward channel*. The other channel transmits from the mobile consumer to the cell site. This is called the *reverse channel*. The forward channel is continually broadcasting to establish its presence and location. Information being broadcast includes the identification of the channel (i.e., frequency, network, modulation); location of the radio transceiver from which it is broadcasting, and management controls to assist in the transmission of signals or radio waves. The forward channel is always hopping from frequency to frequency. This allows it to extend the reach of its broadcast and avoid monopolizing any one frequency.

Once information, or mobile data, is propelled from the mobile end station into the air, the radio waves travel to a radio system within the cell site. Cell sites may have radio systems which are tuned to CSC (i.e. FM modulation) or CDPD (i.e. GSMK modulation). The radio system will receive and process signals which it is tuned for. Signals which are sent using FM modulation are sent to the CSC infrastructure (Figure 2.13 CSC Infrastructure).

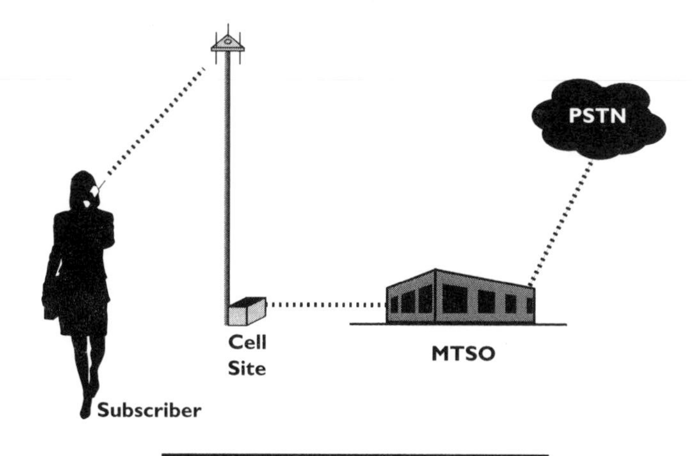

Figure 2.13 CSC Infrastructure.

CSC signals are sent from the cell site to the mobile telephone switching office (MTSO). The designated recipient is determined and the signal is then sent to the public switched telephone network (PSTN). Signals which are sent using GSMK modulation are sent to the CDPD infrastructure (Figure 2.14).

The signal, or mobile data, is then routed to a mobile data base station (MDBS) to the mobile data interexchange switch (MDIS). The MDIS sends the mobile data to a router which then sends it to its designated recipient. The designated recipient could be a host computer or another network (i.e. PSTN or 56 KBps connection).

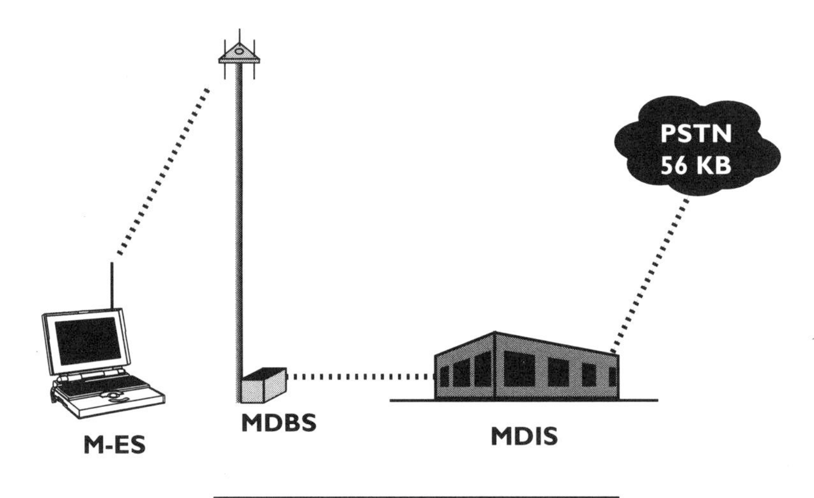

Figure 2.14 CDPD Infrastructure.

Whether the mobile data is sent from a radio modem over the CSC or the CDPD infrastructure, it ultimately arrives at the designated recipient. At this point it is processed and made available to be acted upon.

Providing Ubiquity

Skycell, a service provided by American Mobile Satellite Corporation, will be used to increase the coverage of cellular networks. Having allied with the vast majority of U.S. cellular carriers, Skycell will extend the reach of cellular networks by allowing subscribers to access their networks from anywhere in the United States and its major territories. The Skycell network is scheduled to be fully operational by the end of 1995.

Packet Radio Networks

Distribution of Licenses

As is the case with many networks which provide wireless data communications, packet radio networks generally transmit over the land mobile radio band of frequencies. A network is considered packet radio if it uses packet data technology to transmit information over radio frequencies. Most packet radio networks operate in the 800–900 MHz frequency ranges. Although a few private networks exist which employ packet data technology, there are two major commercial packet radio networks—ARDIS and RAM Mobile Data. These networks will be reviewed later in this Chapter.

Packet Radio Technology

Packet radio networks, or wireless packet data networks, are designed explicitly for the transmission of mobile data. They employ a cell design for the distribution of radio systems (which send and receive mobile data). By dividing mobile data into packets, or discrete pieces of information, these networks are able to transmit mobile data from several subscribers over the same frequencies. Like other networks which employ packet data technology, packet radio networks handle the transmission of mobile data with a high degree of efficiency.

Packet radio networks provide a reliable and proficient means of sending data files or elements. These networks are designed to handle a moderate amount of data. It is generally not economical to send large files (i.e., more than 50kbytes) over packet radio networks. Large files are transmitted more efficiently over networks which provide a dedicated path for communications (i.e., circuit switched networks). To understand why this is so, and how packet data works, it is helpful to compare it to transporting a load of gravel across a river.

Many building projects require gravel to complete the construction of an edifice. Without the proper type and quantity of gravel, the edifice cannot be completed or occupied. Likewise, many reports or computer models cannot produce the required results without the right information. Let's assume that the gravel is in a pile on the bank of a river. There are only two ways to move the gravel across the river. It can be loaded on a truck and driven across a bridge or it can be thrown from one side to the other. The bridge is several miles away, however, and the small size of the pile of gravel makes you consider throwing the rocks to the site which is in view. This would be a preferable method of moving the rocks from one side of the river to another if:

- Your arm or the projectory equipment available to you, has the power and aim to jettison the rocks to the desired location
- You are able to determine whether the rocks arrive on the other side
- You can provide replacement rocks for those which are lost

If the size of your hands, or the holder in the projectory equipment, allows you to move small piles of rocks with continuous thrusts, they will land on the other side of the river in a rather predictable time frame. Moving small groups or packets of information in continuous thrusts, or bursts, is what takes place in *bursty transmissions*.

The wireless transmission of data packets is analogous to throwing or projecting rocks across a river. If you had suitable projectory equipment, a small pile of rocks and someone on the other side who will wave when a sufficient quantity has been accumulated, you may elect to throw the rocks across the river. When you are transmitting a moderate amount of mobile data, i.e., the size of the files are small, packet radio is a viable option. On the other hand, if the pile of rocks was extremely large, you would never even consider throwing them. You would load them onto a truck and drive across the bridge. Likewise, if you are transmitting a large amount of data, a circuit switched wireless or landline network is the best option.

Packet radio networks send mobile data in bursty transmissions. The wireless network provides the information holder, projectory equipment, and the aim that is required to send mobile data to remote locations. These tools are attained according to the capabilities of the radio modem on the mobile device, the proximity of radio systems, and the power which supplies these components. The network works hand-in-hand with communications software to provide assurance, or acknowledgement, that the mobile data has arrived complete and intact. The network and software also take the steps necessary to ensure that any missing or damaged packets are replaced or re-transmitted.

To understand how packet radio networks are designed and deployed, I shall review the two major, commercial packet radio carriers—ARDIS and RAM Mobile Data.

ARDIS

ARDIS, formerly known as Advanced Radio Data Information Services, was established in 1990 to provide commercial wireless data communication services. At that time it absorbed the IBM Network which began operation in 1984 and the Motorola Data Radio Network which began operation in 1986. In its early years, ARDIS focused on providing wireless data communications to IBM's field service technicians and other similar corporate applications. In 1994, Motorola bought out IBM and ARDIS became a wholly owned subsidiary of Motorola.

As in most cell-based networks, the ARDIS radio base stations are placed at or near the center of each cell. The tower locations are selected to provide broad coverage and maximize in-building penetration while minimizing the number of sites. The cells overlap to improve the probability that the mobile consumer will be "in coverage" even while attempting to communicate from deep within a building (see Figure2.15).

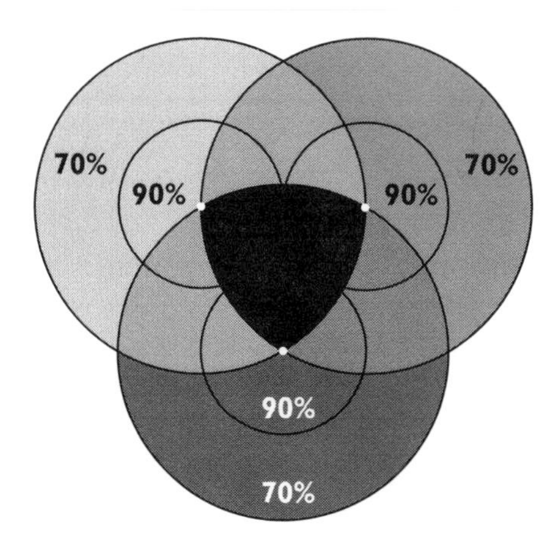

Tower location to maximize in building penetration with minimal number of sites (Delco connections)

Figure 2.15 Overlapping cells.

ARDIS transmits over several of the 600 forward channels in the 806-825 MHz radio band of its 410 metropolitan areas. Each frequency is 45 MHz apart. The reverse channels lie in the 851-870 MHz frequency range. All channels (i.e. transmit and receive) are 25 kHz wide. The technologies and placement of radio towers in the ARDIS network allows mobile consumers to send data 10 to 15 miles within their line-of-sight.

ARDIS uses one of two protocols to identify mobile data transmissions - the Motorola Data Communications (MDC) protocol and the Radio Data Link Access Protocol (RDLAP) protocol. MDC operates at 4800 bps and RDLAP operates at 19,200 bps. MDC can group, or packetize, mobile data into packets of 256 bytes or less. This protocol modulates mobile data using the single level minimum shift keyed (MSK) modulation technique. The RDLAP protocol packetizes mobile data at up to 512 bytes. The RDLAP protocol modulates mobile data using a four level frequency shift key (FSK) modulation technique. The ARDIS network infrastructure was originally deployed using the MDC protocol. All capacity expansion since 1994 has used the RDLAP protocol.

To explain the design of the ARDIS network, I shall describe the components as they perform the flow of information, or mobile data (Figure 2.16). An ARDIS subscriber communicates using a computing device equipped with a radio modem, or with a radio modem equipped with an LCD (liquid crystal display) display and keypad. Mobile data travels from the computing device to the radio modem which projects it into the air in the form of radio waves. The radio waves are sent over a defined frequency using the MSK or FSK modulation technique. The radio waves are received by all radio systems within range. All radio towers are connected to a base station which tunes each modulation technique with the same set of frequencies. The radio stations achieve a high level of in-building penetration through *multi-frequency re-use* technique (see Figure 2.16).

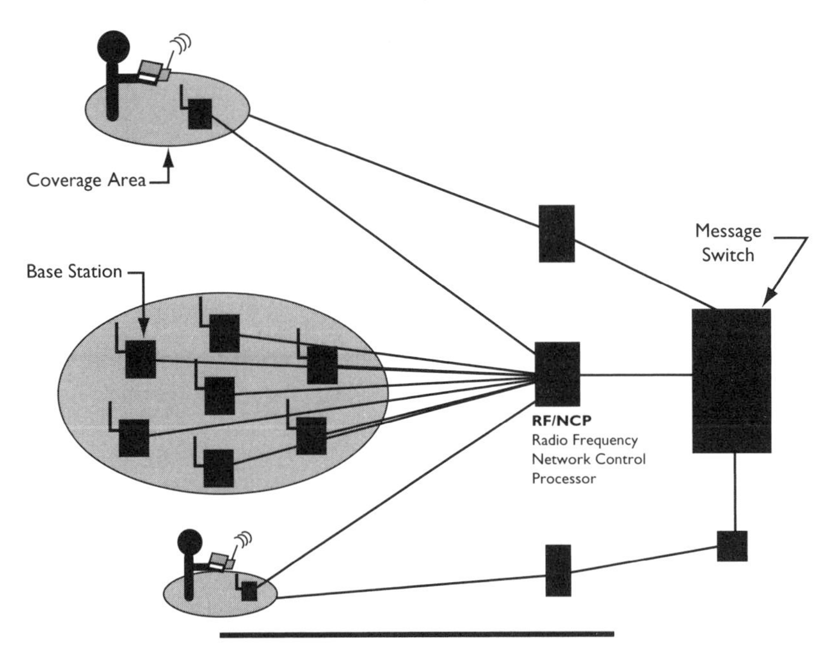

Figure 2.16 The ARDIS Network.

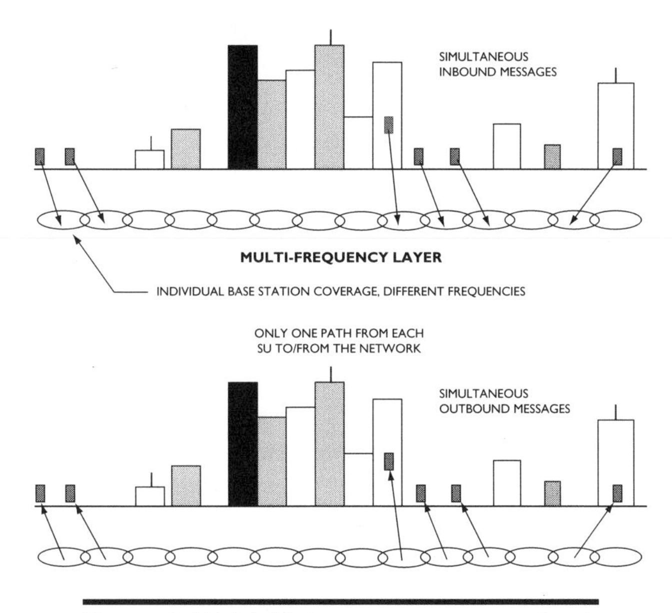

Figure 2.17 Multi-frequency re-use chart.

This technique allows the ARDIS network to send inbound and outbound messages simultaneously. Each message, or radio signal, is sent from the mobile consumers to several radio towers. The clearest message, or the strongest radio signal, is accepted by the network infrastructure Likewise, when messages are sent to the mobile consumer from a radio tower, the network selects the radio tower which will deliver the strongest signal or message.

Each radio base station is physically connected to a Radio Frequency Network Control Processor (RF/NCP) using dedicated lines. The dedicated lines are primarily digital circuits operating at 56 kbps. The RF/NCP verifies whether or not the radio signal, or mobile data message, is being sent from a subscriber who is authorized to use the network. If the authorization is approved, the mobile data message is sent to a *Message Switch*.

Each Message Switch is fully redundant and performs several operations and management functions. Subscriber usage and invoices are prepared with the accounting and billing functions. Network traffic and performance is tracked and adjusted with the management functions. Client host integration with the network is achieved with communications software and physical connections to the Message Switch. End-to-end transmission of mobile data from the subscriber to their host computer is enabled by protocol conversion that also takes place on the Message Switch.

From the Message Switch mobile data flows back to a RF/NCP, to a radio station, and then on to the designated recipient, or mobile consumer. If the mobile data is being sent to a host computer, it is sent from the Message Switch through a physical and software connection (i.e., LU 6.2, X.25) to the host computer. The flow of mobile data ends its journey having reached at another mobile consumer or the host system.

RAM Mobile Data

RAM Mobile Data was established in 1988 by RAM Broadcasting Corporation to provide commercial wireless data communication services. At that time RAM Broadcasting Corporation. had nearly 20 years of experience in mobile communications, including paging and cellular phone services. Bell South Enterprises purchased a 49% stake in RAM Mobile Data in 1992. (Bell South is the largest of Regional Bell Operating Companies and also provides paging and cellular phone services.)

Using a cell-based network layout, RAM's radio base stations (Figure 2.18) are placed in locations that optimize network coverage and capacity. Cells are structured so that they overlap each other, forming the theoretical hexagonal shapes of coverage (see Cellular and SMR). The frequency re-use technique (see SMR) is used to maximize traffic capacity by using all available frequencies.

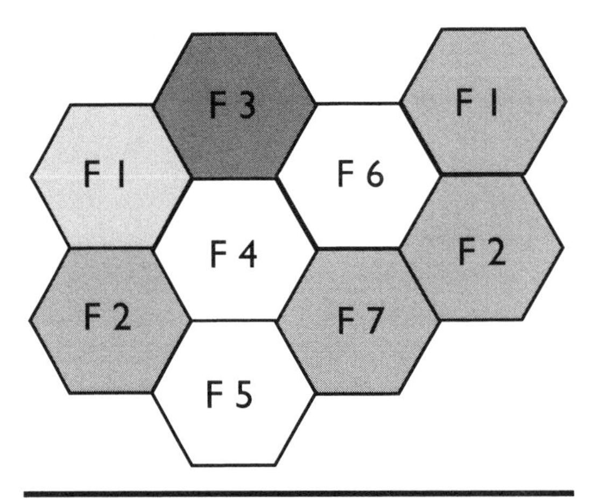

Figure 2.18 The hexagonal shaped cells.

RAM utilizes the 200 SMR channels in the 896–901 MHz and 935–940 MHz frequency ranges. Coverage is optimized by allocating 10 to 30 channels from this range in each city or Metropolitan Statistical Area (MSA). Time division multiplexing (TDM) is used to send mobile data from different sources alternately in the same channel. Each channel is 12.5 kHz wide and the transmitting and receiving frequencies are separated by 39 MHz. RAM uses the Mobitex Asynchronous Communication (MASC) protocol to control the interface between the computing device and the radio systems. Mobitex is considered by many to be a *de facto* standard since it has been used since 1984 by many networks worldwide. Although some countries operate different implementations of Mobitex, the core technology, including MASC, remains the same. MASC performs several transmission management functions on the remote side including initialization of the radio connection, shutdown of the modem, signal strength requests, and other network parameters. MASC modems (Figure 2-19) are also frequency-agile which means they can move from one frequency to another until they find the best available channel for each packet.

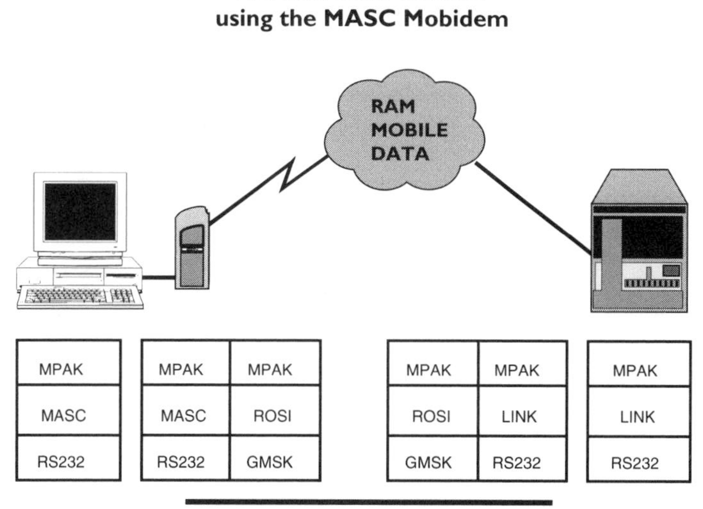

Figure 2.19 MASC Mobidem.

MASC groups mobile data into packets of up to 512 bytes. An additional 11 bytes are used in each packet to manage its transmission (sender ID, addressee ID, network time stamp, etc.). The mobile packet format (MPAK) uses three methods to enhance the transmission of data packets: Forward Error Correction

(FEC), interleaving and automatic repetition request (ARQ). FEC applies extra parity bits (in this case 4 parity bits for every 8 bits of user data) to correct any one bit in error. Interleaving is a process that mixes the bits within a 240 bit block of data so that bit errors caused by signal facing are spread among the smaller FEC codewords. Both the FEC and interleaving provide an added measure of security. ARQ adds two bytes of cyclic redundancy check (CRC) to each block to detect virtually all uncorrected errors so that only those blocks with errors are retransmitted.

Packets are transmitted over the radio channel at a speed of 8 kbps. Radio waves are modulated using the GSMK (Global Systems for Mobile Communications keyed) technique(see Chapter 7: Protocols, Table 7.1). The technologies and placement of radio towers in RAM's network allows mobile consumers to tend data up to 10 miles (Figure 2.20).

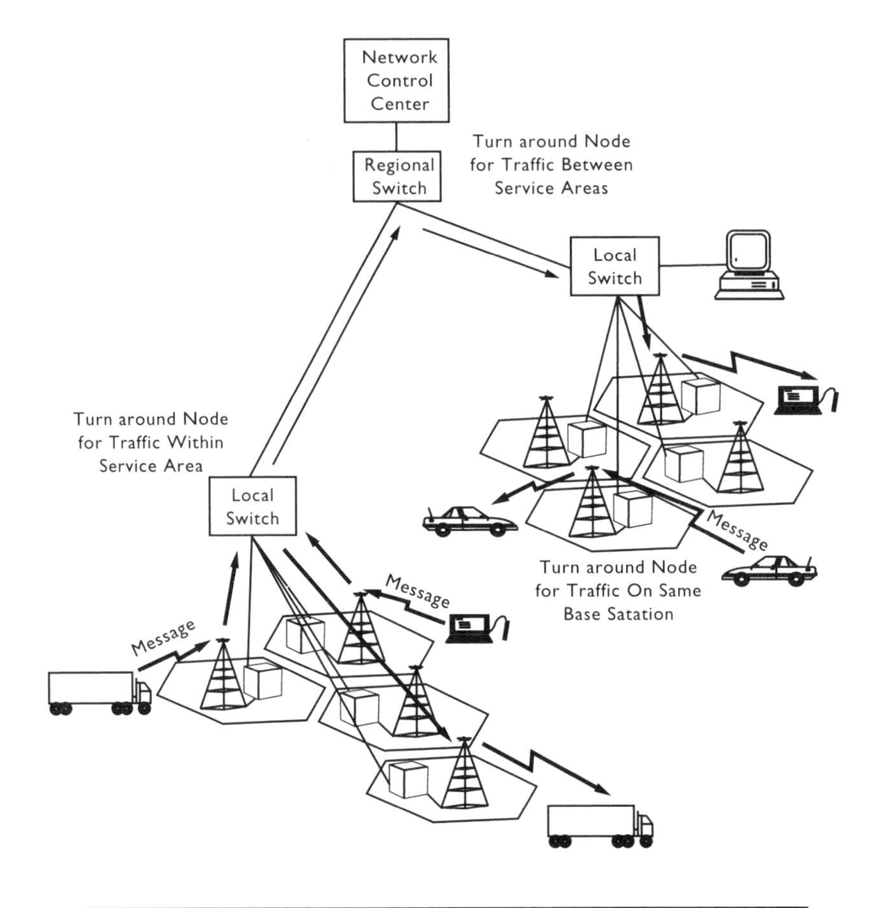

Figure 2.20 RAM Mobile Data Message Transmission Path.

To understand the components of the RAM networks, one should be familiar with the role they play in the flow of information. A RAM subscriber or mobile consumer also communicates using a computing device equipped with a radio modem or with a radio modem, equipped with an LCD display and keypad. Each radio has a mobile access number (MAN) which lets the network know which mobile consumer it is registered to. Mobile data travels from the computing device to the radio modem which projects it into the air in the form of radio waves. The radio modem modulates the radio waves using the GSMK technique. The radio waves are sent over the best available frequency, which is selected based on information from the network's local switch (i.e., signal strength a measure of the ability to receive a reaio wave or signal). The radio modem's ability to communicate with the network allows it to let the network know which radio system it is within range of. This allows the network to perform *transparent and seamless roaming*. Since the radio modem registers its location with the network—no matter where it is—it is not necessary for the mobile consumer to perform this notification. It is seamless because the network is continually aware of exactly which base station the radio modem is in range of. It is transparent because the mobile consumer does not have to invoke any functions.

The radio waves are received by the radio system (i.e. radio tower and base station) within their range which is tuned to their own frequency. The mobile data is accepted by the base station if the modem's MAN receives authorization clearance. If a radio modem has been suspended or terminated for any reason (i.e., theft or delinquent account) authorization may not be given. If the authorization is granted and the MAN for the designated recipient is not registered with the same base station, the receiving base station forwards the packet to its local switch.

The network is constructed in a hierarchical fashion. Each radio system is physically connected to a local switch via dedicated, leased lines. If the designated recipient of the mobile data is within range of another radio system connected to the local switch, the mobile data will be sent or switched to that radio system. The mobile data is then transmitted from the radio system, via radio waves, to the radio modem which is registered to the mobile consumer. The mobile data is then converted (i.e., MPAK to MASC to RS232 format) and sent to the portable computing device. The software works in concert with the portable hardware to display the data for the mobile consumer. If the radio modem has an LCD (liquid crystal display), the mobile data can be displayed on it.

Keep in mind that RAM has a hierarchical network. The local switch is connected to a regional switch which is in turn connected to a national switch with an associated network control center (NCC). The national switch is at the top level of the switching hierarchy. All connections are via dedicated digital lines which are backed up with dial-in lines. Each level in the hierarchy controls a larger number of radio systems, meaning that it governs a larger coverage area. Mobile data will continue to climb the hierarchy from the radio system until it reaches a level which controls the area where the designated recipient is located. Redundancy is provided throughout the system since the mobile data path can be switched to avoid components (i.e., switches) that are not in operation. Additionally, base stations have overlapping layouts to provide partial backup. The designated recipient can be another mobile consumer, a host computer, or any point within a connected private or public network.

The designated recipient can be connected to a base station or switch through one of three connections—a *radio link*, *leased line*, or *X.25*. A radio link supports transmissions of 8 kbps and is generally used by designated recipients who send and receive small amounts of data (e.g., mobile consumers). Leased lines support transmissions of 9.6 kbps to 56 kbps. RAM supports the X.25 and TCP/IP protocols (see Chapter 7: Protocols, Table 7.1) and provides a digital service unit for connectivity. Host computers are often connected via leased lines. Finally, RAM will provide a front-end processor which handles the 2 or 5 step conversion process from X.25 or TCP/IP to MPAK. This gateway, known as Mobigate, provides an X.25 connection between the RAM network and a host computer, private or public network.

Once the mobile data travels from the local or regional switch to the designated recipient, its journey over the RAM network is complete. All packet re-transmissions or acknowledgements take place between the radio modem and the network infrastructure (i.e., local or regional switch). The flow of information, or mobile data, is complete after it has been viewed by the mobile consumer or accepted into an application on the host computer.

Specialized Mobile Radio (SMR) Networks

Distribution of SMR Licenses

Prior to 1993, SMR licenses had been widely distributed among individuals and companies who used them to provide local and regional wireless services. As the market for wireless communications began to grow, these licenses were sold. Major acquisitions and mergers also took place. Cencall changed its name to OneComm Corporation. Nextel acquired OneComm Corporation and Dial Page, Inc., which gave them a substantial number of SMR licenses. Additional SMR licenses were purchased from Motorola. By the end of 1995, the majority of all SMR licenses will be owned by Nextel. This will position Nextel as the first company to offer nation-wide, digital, integrated voice and data services over a wireless network accessible with a single device.

Geotek Communications, Inc., however, presents a formidable challenge as it deploys it GeoNetTM digital, wireless network. GeoNetTM will employ a frequency hopping multiple access (FHMA) radio architecture to deliver integrated voice and data services. With private network operations and interests in North America and Europe, Geotek provides wireless communications services internationally. Its operations include National Band Three (Europe), Gandalf Mobile Systems, Inc., Bogen Communications, Inc. and Speech Design.

SMR Technology

Specialized mobile radio, or SMR, is the name given to the service that is provided over the wireless frequencies that have been used for years by public service organizations (i.e., police, fire, taxi, etc.). These frequencies are commonly known to provide two-way radio services. SMR carriers have provided voice services including subscriber to subscriber communications and voice broadcasting. SMRs operate by sharing the use of narrow-spectrum channels using time division multiple access (TDMA). Transmissions are sent over frequencies in the 800–900 MHz range.

As a result of the utilization of new technologies in many SMR networks, the upgraded newer networks are often referred to as Enhanced Specialized Mobile Radio or ESMR. The essential difference between SMR and ESMR is that the latter uses digital transmission. This allows more robust features such as faster access, consistent sound quality, and a higher level of security. Since we are accustomed in everyday life to speaking and hearing voices in fluid, almost melodic movements, it is easier to listen to analog transmission of voice. Analog transmission more closely approximates the actual sound of our voices. The flow of the voice is captured in the curvature of analog waves. Digital transmission, on the other hand, is accomplished with clear and distinct movements. While the coordination of digital signals is performed with precision accuracy, it requires substantial bandwidth to match the quality of analog voice signals. Wireless networks that handle both digital data and voice transmissions must balance the amount of bandwidth used for each to provide an optimum quality level (Figure 2.21).

Figure 2.21 Analog tramission and digital transmission.

Although the fidelity, or richness, of the digital voice is somewhat poorer than that of the analog voice, the functional capabilities of digital voice- and data-formatting are of a higher order than those available in analog formats.

As with all radio transmission services, SMR uses radio towers to capture radio waves transmitted over the frequencies they are tuned to. The radio waves are sent from a portable device equipped with a transmitter. The radio towers are tuned to recognize and receive the signals being sent to the designated SMR network. In addition to their operating frequencies, SMR networks are differentiated by the manner in which radio waves are received and processed.

Since Nextel owns the majority of SMR licenses, my explanation of the receipt and processing of these radio waves will focus on iDEN—the system employed by Nextel. Integrated Dispatch Enhanced Network, iDEN, is a system developed by Motorola.

Network Design and Components

The technology in the iDEN system supports circuit and packet switched transmission of digital voice and data over the 800 MHz radio band. As a fully digital system handling voice and data, the analog signals which bring voice traffic to the Nextel network must be converted into digital format. This is called *voice coding* or *vocoding*. To accomplish this, pulse code modulation is used to convert analog voice into a 64 kbps digital format. The analog-to-digital converter creates a sample that has 8 bits of information about the voice signal. Every second 8,000 samples are created. This is how the 64,000 bit per second data rate is achieved.

Once the voice transmission has been digitized, the data stream that carries it is compressed so that multiple transmissions can fit into a single radio frequency channel. A coding process is used to reduce the digitized voice signal from 64 kbps to a transmittable signal of 7.2 kbps.

A digital signal, or stream of electric current, must be prepared to be accessed by the radio transceiver and then modulated. iDEN uses a modulation process named M16QAM which permits the transmission of 64 kbps over a single 25 kHz radio frequency channel. This modulation technology which incorporates this speech compression scheme, permits 6 communication paths over one 25 kHz radio frequency channel in the 800 MHz radio band. To minimize interference and reduce errors, the signal is structured to include an *error correction routine* and *guard time*. Error correction allows a faded digital signal to be reconstructed to improve its quality. Guard time gives the transmitter enough time to receive and release power (i.e., power up and power down). Before the signal can be modulated the RF channel must be divided to handle the six communications paths.

Time Division Multiple Access (TDMA) is employed to divide the 64 kbps channel into six different, repeating time slots. Each transmission is assigned to one of these short slots. Remember, the radio transceiver is tuned to recognize the modulation technique as well as the frequency. When the timing is right the radio will provide power to its transmitter, modulate the signal, create a radio wave, propel it through the antenna, and release power. This all takes place during a single time slot.

The technologies employed in the iDEN system facilitate the communication of voice and data over the 800 MHz radio frequency band. Voice coding, advanced modulation, signal structure, and access methods combine to provide clear transmissions while maximizing the volume of traffic handled by the Nextel network.

The design of the Nextel network is similar in structure to that of cellular networks. Geographical coverage areas are divided into sub-areas known as cells. Network design, therefore, is based upon the construction and operation of the cells.

Cells are constructed to maximize the distance that a functional radio wave can travel. The theoretical shape of a cell is hexagonal; however, the shape will often change to enhance communications. Cells overlap to provide high-quality, contiguous coverage. The transceiver, which is usually mounted on a radio tower, is placed near the center of the cell. Directional antennas are connected to each transceiver. The antennas are positioned to enable reception throughout the cell. The reception range of each antenna ends at the edge of the cell. The coverage area for each antenna is called a *sector*.

The distance that a functional radio wave can travel is dependent upon several factors, including the propagation characteristics of the frequencies used, and the topography of the area. Radio waves are designed to travel proportionately outward from the antenna in all directions. Since network components cannot maintain total control over radio waves, the dispersion or propagation of the waves is not always equally proportional. The distance that a radio wave can travel is determined, to some degree, on the manner in which they are dispersed. Hills, valleys, and edifices obstruct the path of radio waves. These geographic considerations help to determine the shape of cells. Sides of the cells may be angled or moved to compensate for interference from land masses or for the propagation of radio waves. The shape and location of the cells helps determine the distance that radio waves can travel.

A transceiver is placed at or near the center of each cell. The directional antennas are positioned to yield three sectors within the cell by extending the coverage area to the sides of the cell. If the antennas cannot achieve complete coverage within the cell due to the topography of the area, the shape of the cell must be modified. The transceiver is part of a system which manages the group of frequencies and channels that are assigned to it. Each frequency may contain several channels to increase the amount of mobile information that can be transmitted over it. One of the channels within each frequency has the job of sending the management information that is required to maintain system operations. This is called the *control channel*. Information that flows over the control channel includes the available frequencies, the identification of subscribers with current connections, and commands being sent by the network. The channels within each frequency are assigned based on a standard pattern. Since cells overlap, all adjacent cells manage a different group of frequencies. The distribution of frequencies and the size or density of the channels are specially assigned to minimize interference and fulfill network traffic volume estimates.

The function of the transceiver is both to receive and transmit mobile data (or voice for the iDEN system). Since many mobile consumers communicate while in transit, they may move out of range of the transceiver during the course of sending or receiving mobile data. In order to maintain the communications link as the mobile consumer moves from location to location, the network uses a procedure called *call handover*. The process of preparing for call handover begins when the mobile consumer first makes contact with the network. The portable device that the mobile consumer carries to access the Nextel network will ascertain the frequency to be used for the connection as well as the frequencies which are assigned to adjacent cells. After the connection has been made, this device continuously monitors the channels and frequencies to determine which one can best handle the communications link when the mobile consumer moves out of range. This information is sent to the network monitoring and control system for processing. As the mobile consumer moves out of range, the network will decide when, and to which cell, the communications link should be transferred. The network then instructs the portable device to transfer, or hand over, the communications link. The actual handover is made by the portable device. As the mobile consumer moves from cell to cell, the network continuously monitors the quality of the radio frequency link to determine when and where additional handovers should be made.

Since cells overlap and traffic volumes may vary, there are often open frequencies which may be used. To increase the volume of traffic that the network can handle at any given point in time, the frequency re-use technique is employed. This technique allows the network to use a certain group of frequencies again and again throughout a group of cells. To implement frequency re-use a formation of cells must be determined that provides the maximum division between cells that use the same frequencies. A network design may employ a cell formation that ranges from a small to large set of frequencies (Figure 2.22).

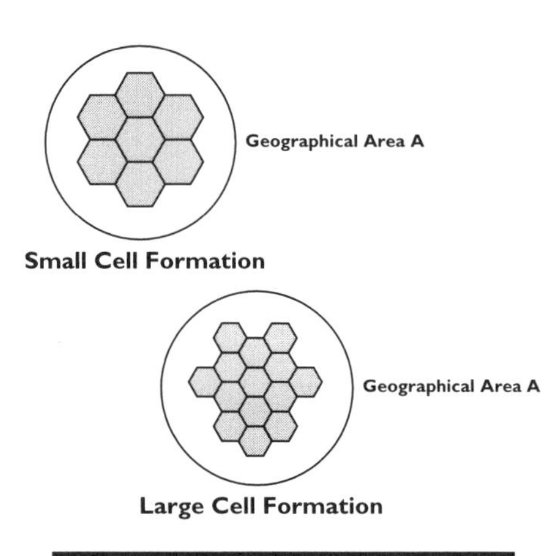

Figure 2.22 Cell formation size.

The iDEN system uses a seven cell formation which repeats seven sets of frequencies. The number of cells in a formation affects the network's ability to handle traffic volume and minimize interference. The distance between cells using the same frequencies is less in formations with a small number of cells. This results in an increase in interference from adjacent formations. As the number of cells in a formation increases, the number of channels which can be handled decreases. This results in a lower volume of traffic that can be handled. In selecting the number of cells in a formation, therefore, the system strikes a balance between the traffic capacity and the level of interference (Figure 2.23).

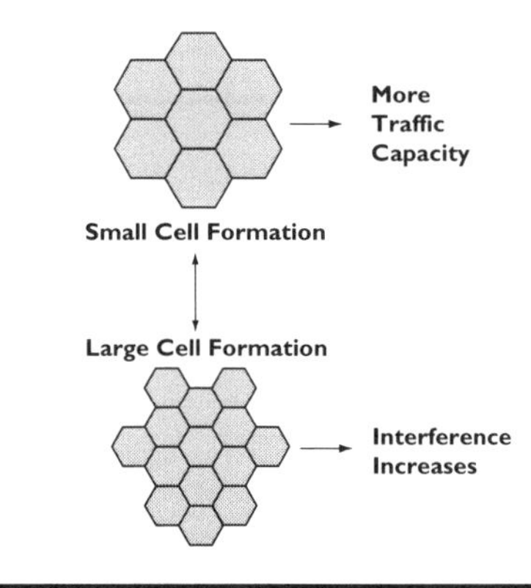

Figure 2.23 Cell formation balance.

Cell sizes are measured by the number of frequencies allocated to it. The maximum size of the cell is based on the distance that functional radio waves can travel. The size of the cell is set by the size of the repeat formation and the number of frequencies available.

$$\text{\# Frequencies in a Cell} = \frac{\text{\# RF Channels Available}}{\text{\# Cells in the Repeat Formation}}$$

The Nextel network uses frequencies in the 800 MHz band. Most cell sites range from 1 to 8 miles. The maximum cell radius is approximately 25 miles.

To understand the components of the Nextel network, let's look at the role that they play in the flow of information. The flow of information begins with the mobile consumer. There are two types of portable devices that can be used to access the Nextel network—a handheld voice and data phone or a vehicle-mounted mobile device. Both units have all the components that are necessary to perform voice and data communications over the Nextel network. These include a radio transceiver, numeric keypad, LCD display, voice receiver, data connection, and an antenna. Each unit also has a unique number that identifies the subscriber to whom it belongs.

The handheld device looks like a cellular flip phone. It has a 1.0-watt transceiver that handles both voice and data. The mobile consumer can use the numeric keypad and voice receiver to use the device as a portable phone. The base of this device has a proprietary data connection. An adaptor is available to convert mobile data to a RS232 serial format. One end of the adaptor plugs into the handheld device; the other end has a 9-pin or 25-pin serial connector. The pins, or physical connectors, determine the type and number of communication functions that are supported in the RS232 format. The serial connector allows a portable computer to be plugged into the handheld device.

The vehicle-mounted mobile device also provides the functionality necessary for dispatch. It has a 3.0-watt dash mount transceiver to handle both voice and data. An external speaker and microphone equip the mobile consumer with voice communications. This device has a RS232 data port for data communications. This allows a portable computing device to be plugged into this device.

The mobile consumer initiates the mobile data flow from a portable computing device. The mobile data travels from the computing device through the data connection on the handheld or mobile device. It is then transmitted from the antenna, in the form of radio waves, to the transceiver in the Nextel network designated for reception. This transceiver is part of what Nextel calls the

Enhanced Base Transceiver System (EBTS). This is the network component that sends and receives transmissions within the cell (i.e. footprint) or geographic area covered and is the central point in the cell (known as the cell site). The EBTS performs several functions including moving or switching the mobile data from one transceiver to another. When the mobile data is designated to be sent to a location that is not accessible by the EBTS, i.e., a distant mobile consumer or to another network, it is sent to what Nextel calls the Base Site Controller (BSC). Each BSC may handle several cell sites.

If the destination of the mobile data is beyond the domain of the BSC, it is sent to a Mobile Switching Center (MSC). The MSC performs several functions including moving the mobile data to another network (i.e., PSTN or cellular) or handing over the call to a remote cell site. The MSC is a pivotal point for the flow of mobile data in the Nextel network. It is at the MSC that mobile data takes the turn towards its final destination. From the MSC, mobile data is sent to another subscriber or to a host computer. If the mobile data is designated to be received by another subscriber on the Nextel network, it will travel from the MSC to a Base Site Controller and on to an Enhanced Base Site Controller. From that point the mobile data travels through the antenna, in the form of radio waves, and is received by the handheld device or mobile device of the intended subscriber. The mobile data then travels to its ultimate destination—the computing device that receives it for processing. If the mobile data is designated to be received by a host computer, it may travel from the MSC to the Public Switched Telephone Network. The host computer may be accessed using a variety of communication links including a modem pool or local area network.

Network Features and Services

As with most wireless networks, the pivotal point is enhanced by the power and functionality of other network components. The MSC is connected to other network components, which provide additional services. These components empower the network to manage the routing and processing of information. They also allow the network to perform additional services such as paging, group-dispatch voice communications, fraud detection, and billing. Nextel promises to deliver additional services to enhance the handling of mobile data in the future. As the network buildout continues and future enhancements are implemented, Nextel's services for wireless computing should increase (Figure 2.24).

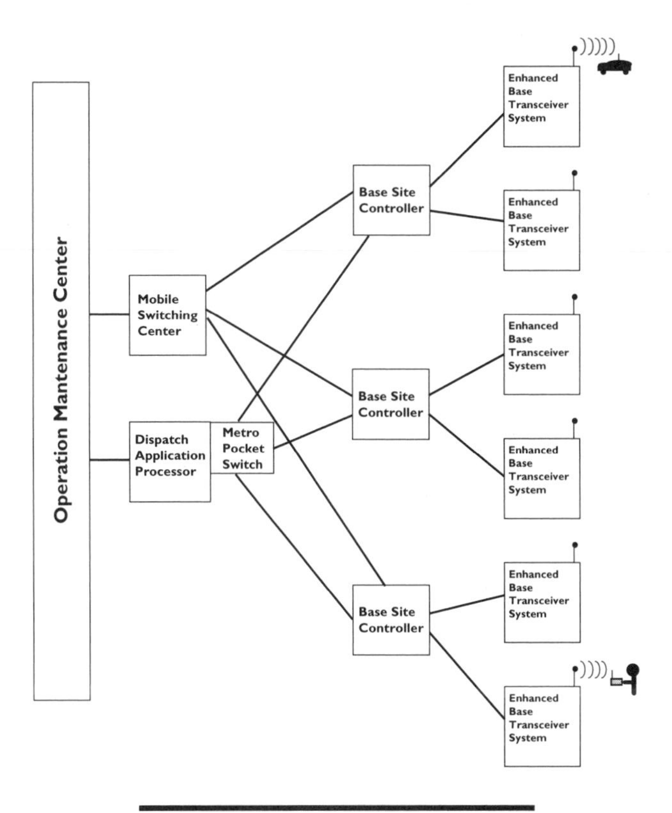

Figure 2.24 The Nextel Network.

Personal Communications Services (PCS)

Distribution of Licenses

Personal Communications Services, or PCS, is a new class of wireless communications that has been made possible by the FCC's reallocation of radio spectrum to new frequencies and license auctions. Microwave systems which operated over targeted high-band frequencies had to be moved. UTAM, Inc., is a non-profit organization which serves as frequency coordinator for unlicensed PCS deployments until all microwave links are relocated. UTAM estimated that 335 microwave licenses must be relocated at a clearing cost of approximately $70 million. In addition to moving microwave systems to complete the reallocation of spectrum, licenses had to be provided to future carriers. For the first time in its history the FCC auctioned off licenses. These activities gave way to a family of PCS services which will be delivered over three bands of the radio spectrum - 900 MHz narrowband PCS, 2 GHz broadband PCS, and unlicensed PCS (1910-1930 MHz). It is expected that services provided by PCS will include advanced paging and messaging.

The availability of unlicensed PCS will allow consumers to own or lease systems or devices for wireless communications. Systems or devices which might include options for this class of service include wireless local-area networks, personal digital assistants, and serial convertors for notebook computers. Since the mobile consumer would purchase or lease equipment, airtime fees will not be incurred. Any product that complies with FCC rules can gain approval to access the unlicensed PCS spectrum. A total of 20 MHz is allocated for this class of service. 10 MHz is provided for *asynchronous* products and 10 MHz is provided for *isochronous* products. The majority of the PCS spectrum, however, is reserved for license holders who will charge for access via their systems. These are the companies who submitted winning bids in the PCS license auctions.

Winners of PCS licenses were awarded the right to transmit over a given set of frequencies within a defined geographical area. The distribution of coverage areas is based upon Rand McNally's 51 major trading areas and 492 basic trading areas. A total of 3,554 narrowband licenses were awarded to provide coverage in one of four levels of service areas—nationwide, regional, major trading, and basic trading. A total of 2,074 broadband licenses (Figure 2.25) were awarded to provide coverage in two levels of service areas: major trading and basic trading.

2 GHz BROADBAND PCS AUCTIONS

FREQUENCY BLOCK	AMOUNT OF SPECTRUM	SERVICE AREA	NUMBER OF LICENSES
A	30 MHz	MTA	51*
B	30 MHz	MTA	51
C	30 MHz	BTA	493
D	10 MHz	BTA	493
E	10 MHz	BTA	493
F	10 MHz	BTA	493

Figure 2.25 2 GHz Broadband PCS Allocation.

National licenses allowed coverage throughout the United States. The frequencies over which nationwide license holders can operate, however, vary for each carrier (see Figure 2.26).

Market Number	Name
N-1 [50-50KHz paired]	Paging Network of Virginia
N-2 [50-50KHz paired]	Paging Network of Virginia
N-3 [50-50KHz paired]	KDM Messaging Company
N-4 [50-50KHz paired]	KDM Messaging Company
N-5 [50-50KHz paired]	Nationwide Wireless Network Corp.
N-6 [50-12.5KHz paired]	Airtouch Paging
N-7 [50-12.5KHz paired]	Bell South Wireless
N-8 [50-12.5KHz paired]	Nationwide Wireless Network Corp.
N-10 [50KHz unpaired]	Paging Network of Virginia
N-10 [50KHz unpaired]	Pagemart II, Inc.

Figure 2.26 Narrowband PCS Nationwide Licenses.

The FCC has imposed limitations on the amount of bandwidth that a single licensee may hold. Regional narrowband licenses were awarded to entities that, in most cases, represented a group of companies. With 1,972 broadband licenses and 2,952 narrowband licenses allocated within the basic trading area alone, the distribution of PCS licenses is likely to be extremely widespread for many years to come.

Licensees must fulfill buildout requirements and deliver service to maintain their licenses. To accomplish this the winners have to identify and negotiate sites, prepare the sites, and install equipment. Nationwide license holders must construct a minimum of 250 base stations within five years and 500 base stations within ten years. Major trading area license holders must provide 25% coverage or construct 25 base stations within 5 years, and provide 50% coverage or construct 50 base stations within ten years. Basic trading area license holders must construct one base station within their area and provide service within one year of obtaining licensing. The goal of the buildout schedule is to attain narrowband coverage for 37.5% of the U.S. population within five years and 75% coverage within ten years of issuing licenses. The final distribution of PCS licenses will depend on the winners' ability to deliver upon the network buildout commitment.

PCS Technology

PCS operates over two areas of the radio spectrum: 2 GHz and 900 MHz. The 2 GHz range is for broadband services. The 900 MHz is for narrowband services. The portion of the 900 MHz range which has been allocated for narrowband PCS includes 901–902 MHz, 930–931 MHz and 940–941 MHz. A 20-MHz piece of the 1900 MHz band, 1910–1930, is provided for unlicensed PCS.

The underlying objective of PCS is to provide wide-area wireless communications for small, inexpensive devices which offer access from within buildings and by users driving along streets and highways. PCS is believed to be able to accomplish this in two ways: its presence and technology. PCS will be the first, viable, direct competitor to cellular communications. As such, it is anticipated that growing competition will result in a decline in prices. PCS, as with many services, will use a cell deign to construct and deploy its network. The major difference between PCS and cellular networks, however, is that cells in PCS networks will be smaller. In the early cellular networks, cells typically had an eight mile radius. Increased network traffic necessitated an increase in the number of cells to maintain performance quality. Cell size in most metropolitan areas is now about a three mile radius. As a matter of fact, micro cells have been constructed in high density areas which have a radius of about 4 miles. Cells in PCS networks, based on the PACS interface, will have radii that range from 5 miles to 0.2 miles. Large cells will have a 1 to 5 mile radius. Small cells will have a 0.2 to 0.3 mile radius.

Smaller cells will allow subscribers to enjoy substantial improvements in mobile communications. In small cells, radio waves do not have to travel as far to reach the radio receiver. That means that less energy is required to send information wirelessly from one point to another. Less energy means low power, which translates into less stringent battery requirements. Not only will batteries last longer, the need for the larger, longer-life batteries will not be as great. Smaller batteries mean smaller, more compact devices. Another advantage of small cells is that there is less chance of geographical interference. Hills, valleys, and other obstructions no longer pose major problems for wireless transmissions over PCS networks because the radio wave will likely reach an antenna before it reaches the geographical obstruction. Small cells also deliver voice quality which is comparable to that of landline telephones. They also enable higher data-transmission rates.

The Pioneer's Preference license for PCS was awarded to Destineer Corporation, according to the FCC, "for having developed and demonstrated the feasibility of significant innovations that will permit delivery of existing and new advanced paging and messaging services in a spectrum-efficient manner." PCS allows four inbound call states: *available*, *screen*, *private*, and *unavailable*. The *available* state allows the mobile consumer to accept all calls except those from telephone numbers that appear on a denied telephone number list which is programmable by the mobile consumer. The *screen* state allows the mobile consumer to view the number and name of the calling party on the screen of the portable device. If the mobile consumer chooses not to answer the call, it will be routed to voice mail or to a designated telephone number. The *private* state allows the mobile consumer to receive only those calls from people on their defined private list. All other calls are routed to voice mail or to a designated telephone number. The *unavailable* state will route all calls to voice mail or a designated telephone number.

The size of cells and level of functionality that will prevail in the PCS network will depend upon the air interfaces that are employed. The degree of interoperability will also be dictated by the selection of air interfaces. The air interface determines how mobile data is transferred between the mobile consumer and the radio base station. That is, the method which is used to access radio waves. Two standards groups are working to gain agreement upon a set of interoperable air interfaces—the Joint Technical Committee of the Telecommunications Industry Association and the Committee T_I which is accredited by the American National Standards Institute. Several air interfaces have been suggested for consideration including:

- IS-54 based TDMA
- IS-95 based CDMA
- DCS based TDMA
- Composite CDMA/TDMA
- PACS TDMA (originally known as WACS)
- DCT based TDMA
- Wideband CDMA
- inFlexion

See the chapter on protocols for an explanation of these air interfaces elsewhere in this book.

Network Structure and Distribution

At least eight interfaces are under consideration by network providers, and designs have not been finalized. Since there are more than 4 companies who hold licenses to provide nationwide coverage, and many more who hold licenses for regional and local coverage, it is unlikely that a single network design will prevail. The basic design for radio network, however, will provide a framework for understanding the configuration of technologies in these networks. PCS will use a cell design with *pico* and *micro* cells. Pico cells have a radius of about 0.25 miles. Micro cells have a radius of about 1.5 miles. As with most cellular network designs, adjacent cells operate on a different set of channels or frequencies. The distribution of frequencies among cells combined with frequency reuse techniques combine to increase the traffic volume that the network can handle (see SMR network design).

The design of the network infrastructure will be based on the basic design of a radio network (see Figure 2.27).

Proprietary interfaces are often used between the switch and the controller. The non-proprietary interfaces have been proposed to provide this connection in PCS networks: the A interface, A+ interface and the C interface. In addition to their design, these interfaces vary by the individual component, or configuration of components, among which they provide connection. The MSC is a configuration of equipment, including a communications switch, that provides the movement of mobile data from the PCS network to other locations (i.e., the host computer or PSTN). The Base Station Controller (BSC) is a configuration of equipment, including a communications controller, which provides network management functions as well as the processing and configuration of the mobile data itself. The A interface will provide the connection between a MSC, which houses the switch, and the BSC, which houses the controller. The North American Cellular A+ interface will provide the connection between a MSC and the controller. The New Generic C interface will provide the connection between the actual switch and controller.

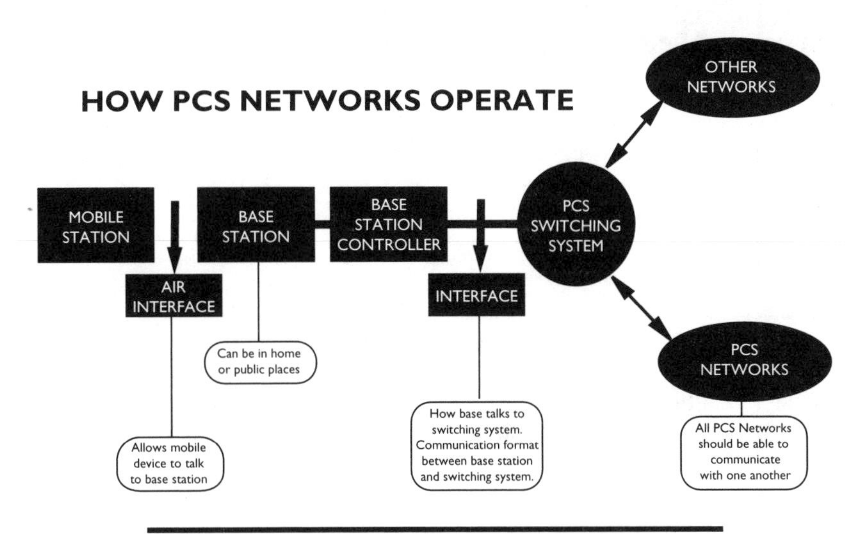

Figure 2.27 Typical PCS network infrastructure.

The air interface is the modulation technique (i.e., CDMA, TDMA) and the method for packaging mobile data so that it will be recognized by the radio system. Remember that a radio system recognizes radio waves by the frequency over which they travel, the modulation technique that was employed before sending them, and the network identifiers and management information packaged with the mobile data before transmittal. Since some PCS carriers will transmit over the same frequencies, the interface will distinguish between pieces of mobile data being sent over different networks. Industry associations are working on defining an interface that may be agreed upon and used by all carriers. Although a small group of popular interfaces may result from these efforts, a common air interface may not emerge. If interoperability is achieved between PCS network infrastructures, however, the lack of a single common air interface may not severely impede communications among PCS networks. Lack of a common air interface will prohibit devices designed to communicate with a given PCS carrier from communicating with other carriers using the same device. This means that mobile consumers may have to change radio modems to communicate in different PCS coverage areas. The alternative to changing modems is to use a more expensive, multimodal access device (e.g., phone and modem).

Paging Networks

Paging networks have been in operation for nearly 50 years. The first page ever sent is believed to have been received in 1950 by a doctor on a golf course. In addition to doctors, millions of people from all walks of life use pagers today. Repair technicians, salespeople, coaches and emergency service personnel use pagers. Many families use pagers to assist in locating their children.

There are several paging networks in operation throughout the United States. Paging networks may provide coverage for defined metropolitan areas or the entire United States. Some carriers also provide service to several other countries. As one of the oldest and more established means of wireless communications, paging networks are usually the most pervasive.

Paging Technology

The technology employed by traditional paging networks supports one-way transmission of mobile data. Although two-way communication is usually mandatory for wireless computing, applications do exist that can be successfully implemented using one-way paging (see Case Study: Spaulding and Slye). It should be noted that two-way paging services will be provided over the emerging Personal Communications Services (PCS) networks.

The most common form of paging uses the POCSAG (Post Office Code Standardization Advisory Group) protocol to handle the transmission of mobile data. While this protocol does not limit the maximum length per message, many carriers limit the size of messages to increase network throughput. Skytel, for example, limits alphabetical characters to 240 per message and numeric characters to 21 per message. POCSAG divides transmissions into eight frames. Each frame contains two thirty bit words which either represent addresses or data. Each pager—the device carried by the mobile consumer—is assigned a specific frame. The pager wakes up to listen for messages when its assigned frame is being transmitted. It remains asleep during the remaining seven frames. Therefore, the pager actively consumes power 1/8 of the time when it is listening. Power consumption is very minimal the remaining 7/8 of the time when it is asleep. This allows the paging device to conserve battery life.

The pager recognizes messages which are sent to it by the *capture code*, commonly known as the *capcode*. The capcode occupies one of the thirty bit words (e.g., data words) in the assigned frame. This is the number which is assigned to the pager device and is often inscribed on the outside of the device.

Data words in the frame carry either five numeric digits as 27/7 alpha characters in twenty information bits per code word. Any number of data words can be strung together to transmit very long messages.

Transmission of mobile data using the POCSAG protocol is facilitated by algorithms which reside in the pager itself. These algorithms allow the pager to avoid multipath interference and also assist the mobile consumer in recognizing incomplete messages. Multipath interference occurs when the same message is transmitted over the same frequency from multiple radio stations. When this occurs in television broadcasts the viewer sees multiple images which are often referred to as ghosts. Modern television technology (i.e. tuners in TV sets and the broadcast TV network infrastructure) have substantially reduced the occurrence of ghosts. When this occurs during paging transmissions, a distortion occurs in the detected message that the pager must successfully overcome to decode the message. An algorithm in the pager device eliminates the effects of multipath interference on messages displayed for the mobile consumer. Once a message has been received, this algorithm ignores all duplicate messages and, therefore, does not display them for the mobile consumer.

Another algorithm helps the mobile consumer recognize incomplete messages. Since traditional paging networks are one-way, the messages do not have *footers*. A footer is a code placed at the end of a data message to indicate its termination. A one-way network does not have a return transmission path through which to acknowledge receipt of the message. It is not possible, therefore, to request retransmission of an incomplete message. A footer, therefore, is of no value. The "end-of-message" algorithm helps to compensate for the lack of a footer. This algorithm displays unique characters which are thought to be understood and also for those which are unrecognizable. Another unique figure is used to denote the end of a message. If an "end-of-message" character appears before the logical end of the information displayed by the message, it is reasonable to assume that the transmission was incomplete. Likewise, if unrecognizable figures appear before the "end-of-message" character, the mobile consumer can safely assume that the transmission was not successful.

Network Design and Components

The flow of mobile data in paging networks begins with the person who is sending a message to the mobile consumer. When a message is entered the sender must enter an identification number which stipulates to whom the message is being sent. This identification number is assigned to a capture code or is the capcode itself. The capcode becomes a field in the header which is used to direct the message to the designated pager device. There are two types of paging which ultimately determine how the capcode is assigned–*selector level paging* and *end-to-end paging*. With selector level paging, each pager is assigned a unique phone number. This number is often the capcode and is printed on the outside of the paging device. With end-to-end paging each mobile consumer or subscriber is assigned a personal identification number (PIN). A main telephone number is used to dial into the paging network. The capcode is either the PIN number or a number which is uniquely assigned to the mobile consumer carrying the pager.

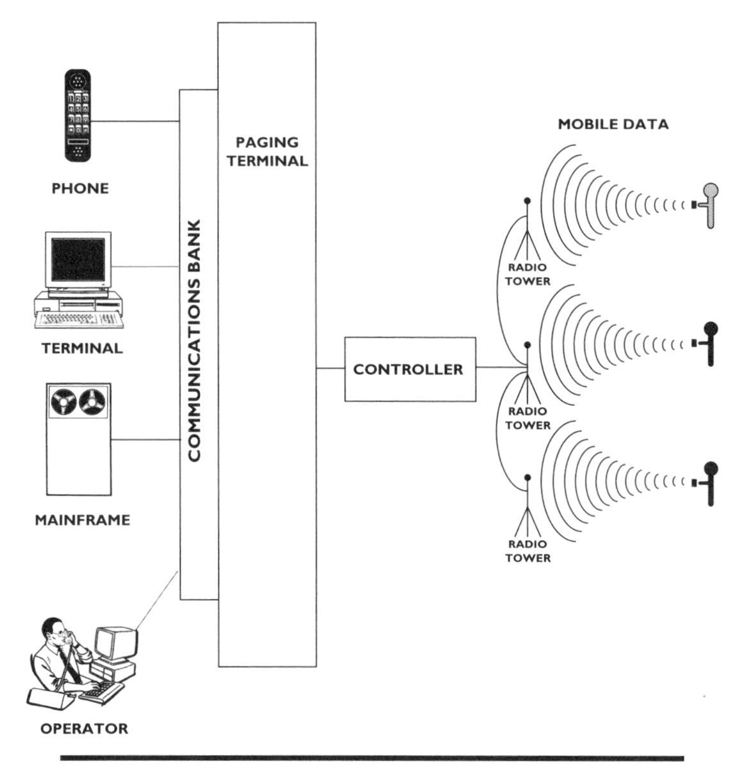

Figure 2.28 Typical paging network infrastructure.

The message may be entered from a telephone, computer terminal or a host computer. (see Figure 2.28). If a telephone is used, the message can be composed by pressing numbers on the keypad or by talking to an operator who, in turn, composes and sends the message using a computer terminal linked to communication lines. A digital or "tone" equipped handset and telephone line are necessary to send messages directly by telephone. When a message is entered using a telephone keypad, an end of message identifier, usually the # key, is pushed to denote the end of the message. Once composed the message is sent from the telephone via physical landline links to a communications bank. The landline link may be a T1 line, the Public Switched Telephone Network (PSTN), a private branch exchange (PBX), or other communications path equipped to handle analog transmission.

Messages can also be initiated from a computer terminal or host computer. A computer terminal may be used by the person who is sending the message or by an operator who verbally receives the message from the sending party. The message is entered via the computer terminal keyboard. It is then prepared for transmission over the communications link, or formatted, by software resident in the computer terminal and equipment (e.g., modems, LANs) connected to the terminal. Once the message is ready for transmission it is sent to the communications bank.

Paging messages can also be initiated form a host computer. The host computer may be a personal computer, minicomputer, or mainframe. It is programmed to send a paging message based on a set of occurrences which warrants its transmission. For example, a portfolio management application may reside on the host computer to track and process the movement of stocks, bonds, or other financial instruments. Individual stocks or a group of instruments may be assigned to a broker or trader who is assigned to track and manage them. If the price of a stock moves above or below a certain level, the program can send a paging message to the designated broker or trader who is carrying a pager. The message can inform the broker or trader of the current price so that they can determine whether the portfolio owner should change their position. That is, whether they should sell the stock or buy more. A host computer can, therefore, be used to send immediate messages when warranted without the need for human intervention. As with other methods of originating paging messages, the host computer sends the message to a communications bank.

The communications bank is physically connected to a Paging Terminal or computer. This is a pivotal point in the transmission of mobile data on a paging network. The paging terminal performs several functions. First of all, it prepares the message to be received by the network. It then sends the message to a controller which is often physically linked to the paging terminal. In some networks the controller may be linked to the paging terminal by satellite. This allows the network to achieve a greater coverage area. The controller is physically connected to the radio system. The radio system includes several radio transmitters which are physically linked together. The links may be via telephone lines, T1 lines, or a control frequency in the 35 MHz radio band. These links manage the performance and communications of each radio transmitter connected to the network. Each transmitter has a base station which handles the conversion of the mobile data into radio waves. The radio transmitters handle multiple frequencies and multiple baud rates. They are equipped to send mobile data from 1200 and 2400 baud pagers. The newest technology permits paging speeds up to 6400 bps. Most of the older 512 baud pagers have been retired. Once the mobile data reaches the radio system it is relayed by each transmitter in the system. This gives the message every possible opportunity to reach mobile consumers in the coverage area. Every transmitter, however, sends over the same frequencies. To avoid the collision of messages, the radio systems broadcast in simulcast. That is, each radio transmitter fires up, or transmits, messages a few milliseconds apart. This ensures that identical messages are not transmitted over the same frequency at the same time. The collision of messages, or crashes, are therefore avoided.

After the message is airborne, the final step in the flow of mobile data takes place. The message must be received by the pager. All pagers have crystals which are tuned to a single frequency. The frequency is assigned to the carrier to whom it is licensed. The carrier also authorizes the mobile consumer to whom the pager is registered. The crystal vibrates when it receives signals on its frequency. (This is not the same vibration that the mobile consumer feels when they receive a message on a vibrating pager.) The crystal is, therefore, the receiver in the paging device.

As I pointed out in the Paging Technology section, each pager is assigned a specific frame. These frames are based on specific and discrete time intervals. It is necessary, therefore, for the pager to remain coordinated with the network to receive information being sent in its designated frame. This is particularly true when the mobile consumer turns on a pager which has been off for a period of time. To maintain time coordination with its assigned pagers, paging networks transmit synch codes periodically so the pager can keep in time with the network. The synch codes let the pager know which of the eight frames that its messages will be transmitted in. The synch codes are received by the crystal in the pager.

In addition to synch codes the crystal will receive all signals which are transmitted over the frequency to which it is tuned. Although the crystal in the pager may receive signals on its frequency, it will not accept the message if its header does not contain the capcode of the paging device. If the message does have the capcode in its header, the pager will accept it and display its contents on the screen for the mobile consumer to see. If the pager is configured to notify the mobile consumer that a message has been received, it will vibrate or emit a sound.

The flow of mobile data in a traditional paging network takes place in one direction only—from the sender to the mobile consumer. Messages are sent via simulcast, which increases the likelihood that the mobile consumer will receive it. The scope of coverage is dependent upon the placement of the radio transmitters. Many radio transmitters are within buildings and some are even under the ground. The capacity of a paging network at any point in time depends on the number of messages being transmitted, the size of the messages, and the speed or baud rate at which the messages are being sent. Higher baud rates allow the transmission of more messages.

Two Way Paging

Two-way paging requires a more robust protocol and greater bandwidth. It will be delivered by most traditional paging carriers via the emerging Personal Communications Services (PCS) networks.

Satellite Networks

Satellite Technology.

Satellites use microwave radio relay technology operating at high frequencies. Microwave radio technology has been widely used for more than 40 years and is extremely stable and reliable. Satellite communications are fast and can handle large amounts of data. They transmit at high frequencies—above 50 MHz. Radio waves at these frequencies can travel almost unlimited distances. They can travel through the ionosphere into space. However, they can not pass through solid earth. Transmissions at these frequencies, therefore, must travel in straight lines, unobstructed, from one point to another. That is, they must be line-of-sight. The circumference of the earth limits a microwave's transmission path to about 30 miles. Rather than build a relay tower every 30 miles, satellites have been launched to capture or receive radio waves traveling at these frequencies and reflect or transmit them to far away locations. Since satellites are positioned miles above the earth's surface, their line-of-sight can range up to 22,238 miles. Satellites, therefore, can provide a communications link between two or more relay or Earth stations which are many miles away from each other. The addition of satellite links is one of the many strategic elements that wireless carriers must consider in their quest for ubiquity (Figure 2.29).

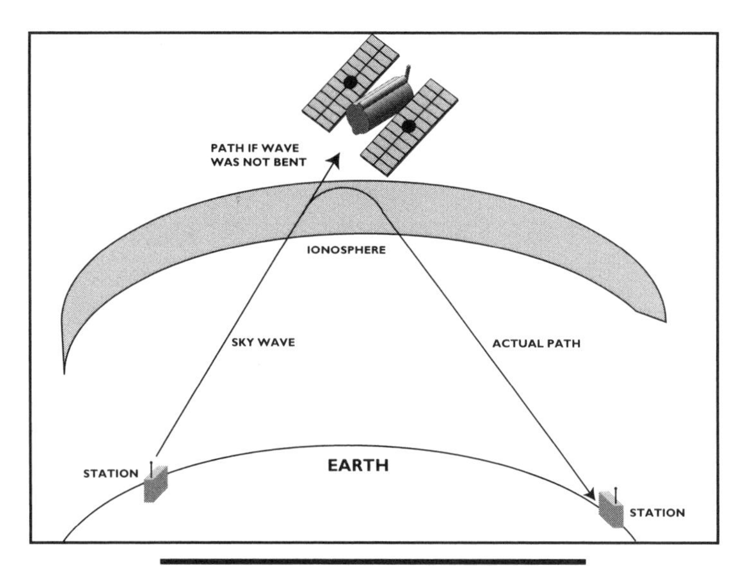

Figure 2.29 Satellite components.

Communications satellites may be used to send information around the world. They can receive microwave signals from one continent and send them to another. Two types of satellites are used for communications: active repeaters and passive repeaters. Passive repeaters do not have transmitting equipment. They send signals by merely reflecting them to a different location. Active repeaters receive signals, amplify them, and beam them to remote locations. Since active repeater satellites may be more appropriate for sending mobile data, the focus of this chapter will be on how they operate.

The primary component of a communications satellite, the one that, handles the transmission of radio waves, is called a transponder or radio repeater. A satellite has several transponders. Each transponder consists of a receiver, a frequency shifter, and a power amplifier. These components provide the intelligence for communicating with relay stations on earth. Relay stations are constructed and positioned to support communications with satellites operating in a certain orbit path. They are also configured to support a group or band of frequencies. Mobile data is sent from the relay station to the satellite over a channel within a designated frequency. The communications link which provides the path for sending signals from a relay station to a satellite is called an uplink and operates at a high frequency level. The transponder's receiver is tuned to a channel within the relay station's band of frequencies. The transponder's receiver accepts the signal that carries the mobile data for processing. Once the signal carrying the mobile data has been received, it is handled by the frequency shifter. This component prepares the signal for transmission back to earth. The communications path which carries the signal from the satellite back down to the relay station is called a downlink. The downlink operates on a lower level of frequency than the uplink. The frequency shifter lowers the signal from the uplink frequency level to the downlink frequency level. This means moving the signal from one transponder to another. Remember, each transponder has its own receiver which is tuned to a specific channel. The satellite downlink is more difficult. Radio waves can lose power, or attenuate, as they travel through the ionosphere and the atmosphere. Attenuation is less likely to occur when radio waves travel at lower frequencies. Once the signal is ready to be sent at a lower level frequency, the power amplifier is engaged to produce enough energy to generate radio waves. The mobile data is then sent with the signal to another relay station on earth. This station can be far away from the transmitting relay station—even on another continent.

Satellites must be launched into orbit before they can perform communications worldwide (Figue 2.30). Geostationary orbits provide a fundamental understanding of how satellites move around the earth. When you put one body which is attracted by gravity (i.e. satellite) around another which has a gravitational pull (i.e. earth), you can set the orbital velocity of the first body such that when it falls it is tangent to earth. Thus, it stays in orbit. When the body is closer to earth it is drawn in by gravitational pull faster. The orbital velocity must be faster to counteract the stronger gravitational pull. Since low earth orbiting satellites (LEOS) are closer to earth they must spin faster. A space shuttle, for example, goes around the earth in less than a day but the moon takes 28 days to orbit the earth. An orbit is measured in hours. At a point between the altitudes of a space shuttle and the moon is an altitude where the orbit is exactly 24 hours. If you place a satellite in this orbit above the earth's equator, it will fall around, or orbit, the earth in one day. This is the geostationary orbit.

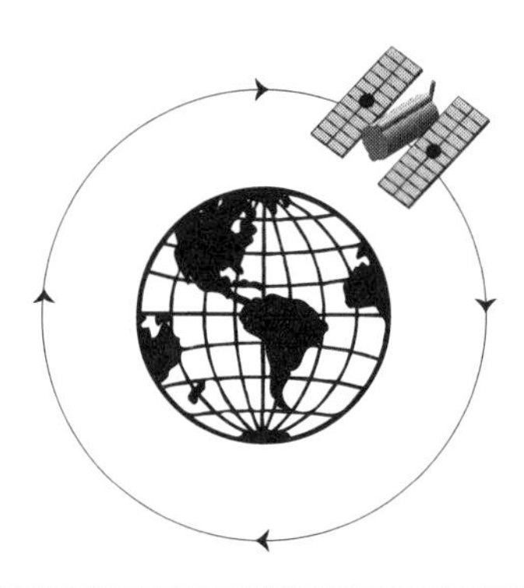

Figure 2.30 Satellite communications.

Once they have been launched, satellites must be stabilized. There are several stabilization techniques including spin-stabilization and three-axis stabilization. Regardless of the technique which is used, satellites must be stabilized in orbit to maintain the number of hours necessary to provide continual coverage. To achieve this the antennas must be positioned to maintain line-of-sight with earth stations. Altitude stabilization is necessary to allow the directional high-gain antennas to be pointed in the desired directions. This improves the reliability of radio wave transmissions.

Since satellites are launched to orbit the earth, those which are in low orbit are within line-of-sight of transmission points for fewer hours than those in higher orbit. Low earth orbiting satellite (LEOS) systems, however, can receive radio waves at lower frequencies and thus require less power. On the other hand, global earth orbiting satellite (GEOS) systems can achieve broader coverage with fewer satellites. GEOS satellites maintain an orbit which is 22,238 miles above the earth's surface.

Due to the broader coverage of satellites, they are often used to extend the reach of wireless network carriers or as the transport facility for vehicle tracking or global positioning (GPSS) systems. The higher cost and greater power requirements of satellites make these roles technically and economically feasible. The cost of developing, building, launching, and operating a satellite can easily reach millions of dollars. A substantial subscriber base is necessary to bring prices down to a level that most consumers are willing to pay.

Wireless mobile computing applications that communicate via satellite directly to the mobile device are usually challenged by portability and cost-justification issues. The nature of the mobile worker's environment and routine must allow the wireless computing device to be portable and operable. This means that all the equipment needed to operate the device, including the radio modem and batteries, must not be too heavy to move from one place to another. The batteries must also last long enough to perform the required functions throughout the workday or shift. If this is not possible, it must be operationally feasible to carry extra batteries or recharging equipment.

Since satellite communications require more power than other wireless networks, batteries to provide this power are larger and heavier. The batteries and equipment required to support direct satellite communications throughout the workday are generally too large and heavy to be easily carried. The power for a portable computing device that uses direct satellite communications, therefore, is usually provided by batteries mounted in a car, truck, or other vehicle. The transmitting device is located in the vehicle or within a radio relay station physically connected to the wireless network. Signals sent from a vehicle are reflected off a satellite and on to a radio relay station within the network. Applications using direct satellite communications must also offer a substantial benefit to the consumer to justify the cost of implementation and operation. Many applications that employ vehicle tracking or global positioning systems bring a high return to the consumer. Such applications include deployment of large vehicle based field forces, stolen car recovery, and aircraft guidance systems.

Satellite communications provide high bandwidth for the transmission of large volumes of information. To take advantage of the larger bandwidth available from satellite transmissions, wireless carriers may use several methods to group circuits together. This is called *multiplexing*. The transmitting device, often a radio modem, then creates signals from the grouped data which are modulated before being sent. Modulation may be achieved by varying the signal's amplitude or frequency. This process also includes information which is used to complete the communications transmission. The signal is then sent to the satellite. A transponder within the satellite amplifies the received signal or complex of signals and changes its frequency. This does not change the information content before it is sent back to another radio relay station on earth. When a satellite can communicate between or access several earth stations at the same time it is said to have *multiple-access capability*. Before the signal is successfully transmitted, the frequency is demodulated and information which aids in the communication is stripped off and used. The signal must then be *ungrouped*, or *demultiplexed* to process the channels being carried. The channels are subsequently directed to the intended location. The wireless network carrier can now forward the information contained in the channels to its designated subscriber(s).

Access Methodologies

As we have learned, the transponder is the main component of a satellite enables communication. Techniques that maximize the performance of the transponder, therefore, will increase the efficiency of satellite transmissions. If a satellite operates in quasi-linear mode, several earth stations can use its transponder simultaneously if the power allocated to each carrier is carefully controlled. *Frequency-division multiple access* (FDMA) accomplishes this by allocating a fixed frequency, bandwidth, and power budget to each station. This method requires demodulation and demultiplexing to take place at the radio relay station. The large, active stations must, therefore, maintain a great deal of multiplexing equipment to process channel traffic. Not only does this prove expensive, more efficient techniques are available to handle heavy traffic.

Time-division multiple access (TDMA) allocates usage of the transponder by giving each station an assigned time slot for sending and receiving information. Although these slots are in very short intervals, they are controlled by a crystal clock to allow frequent repetition. TDMA allows the station exclusive use of the transponder during its time slot, therefore permitting it to maximize performance and efficiency. Multiple stations can also be accessed with a single set of equipment.

Commercialization

With the passage of the Communications Satellite Act in 1962 and the establishment of Comsat Corp. in 1963, satellite communications began to gain commercial status in the United States. Prior to this time, satellites had been used primarily for weather tracking and to support government communications requirements. The 1990s promises the largest growth in satellite deployment since their commercialization began. The launch of several satellites are planned over the next few years by companies striving to gain a share of the market for wireless communications. The projected increase in wireless subscribers, combined with their demand for ubiquity, underlie plans for aggressive satellite deployment.

Spread Spectrum Networks

Spread Spectrum Technology

The technologies that we have discussed thus far—cellular, packet radio, SMR, PCS, paging, and satellites—are used for wide-area communications. They all operate above the 800 MHz radio band. Their range is extensive since the radiowaves produced by these technologies travel up to 30 feet and beyond. *Spread spectrum technology*, on the other hand, is primarily used for short-range, or local-area communications. (I will discuss how spread spectrum technology is used for wide area communications later in this section.) The Federal Communications Commission does not require a license to transmit mobile data over spread spectrum frequencies. Rather, the device which is used to send and receive information must be approved by the FCC. To gain approval the device must transmit using less than 1 watt of power, among other things. Three bands in the radio spectrum have been allocated for spread spectrum communications—902–928 MHz, 2.4–2.4835 GHz and 5.725–5.850 GHz. Radio waves are sent over these frequencies from a transmitting device to a receiving device. Unlike most other radio frequency technologies, however, spread spectrum does not transmit a single signal over a single frequency. It sends mobile data by spreading the signals over a spectrum of frequencies. It is this methodology that gives spread spectrum its name.

Spread spectrum communications was developed during World War II for use by the United States military. To avoid enemy interception, spread spectrum technology was designed to incorporate a high level of security. The radio waves which carry mobile data, or signals, are moved between frequencies using one of four techniques—*chirp*, *time hopping*, *direct sequence* or *frequency hopping*. Direct sequence and frequency hopping are the techniques commonly employed in wireless computing systems. The direct sequence technique sends a signal from one frequency to another along a sequential path. The signal may skip over a frequency, however, it does not move back and forth between frequencies. The pattern of the sequential movement may vary using direct sequence. To intercept a signal being sent using the direct sequence technique, the intruder must decipher the pattern of frequencies in the signal path. The frequency hopping technique sends a signal from one frequency to another in a disparate pattern. The signal "hops" among frequencies in a non-sequential pattern. It can ride on a single frequency several times during the course of its transmission. Interception of a signal being sent using the frequency hopping technique is generally more difficult and virtually impossible with detailed specifications.

Local-Area Wireless Networks

Conventional local area networks physically connect devices (computers and peripheral equipment) with cables and wires. A major benefit of using spread spectrum based local area networks is that equipment can be connected via radio waves. Wires and cable connections are unnecessary. The consumer is only mobile to the extent that he or she can move their computing equipment from one location to another, remaining within range of the network.

The flow of mobile data through local area networks employing spread spectrum technology is relatively straightforward. Mobile data originates in the computer used by the person who is sending the information. The data then travels through a serial port, or expansion slot, to a network connector or interface card. The network software packages the identifier of the sender and that of the designated recipient with the mobile data. The network connector contains a transceiver which converts the electric current into a radio frequency signal or radio wave. The radio wave is propelled through an antenna which is externally mounted on the network connector or interface card. The radio waves travel up to 30 miles, reaching all network nodes or access points within its range. The radio wave is received by the designated network node when its identifier, which is included with the mobile data is recognized. The network connector or interface card then translates the radio wave back into electric current. It is then processed by the computer or other equipment connected to the network.

Spread spectrum-based local-area networks have grown in popularity as consumers find them both functional and cost effective. Several companies now deliver products with this technology.

Wide-Area Spread Spectrum

Spread spectrum technology has also surfaced in wide-area networks that currently provide coverage in limited geographical areas. These networks operate in the 902–928 MHz range and generally support transmissions of up to 30 miles. Mobile data is sent at speeds of up to 100 kbps. A cell design is used by these networks with microcells that can handle up to 163 channels and 200 transmissions, or mobile consumers, within each cell site. The micro cells are physically connected via cables or other communication lines. These connections are the backbone of the network, often supporting gateways for communications with other networks and host computers. Wide-area spread spectrum networks offer a low-cost, reliable means of providing wireless communications within a defined geographical area. The coverage areas may increase as these networks establish themselves. Although the cost of buildout is lower than most other wide area wireless networks, users of this technology still face the same challenge of negotiating site locations for radio equipment.

Infrared Communications

Infrared Technology

All of the wireless technologies that have been presented thus far use radio waves to carry mobile data. Infrared (IR) technology, however, transmits mobile data through light waves. Infrared transceivers convert a stream of electric current carrying data composed of pulses, or 1s, and non-pulse, or 0s, into light waves and beam them to an infrared receiver. Infrared technology beams light waves using one of two techniques—*line-of-sight* or *diffused propagation.* Line of sight transmission, sometimes described as point and shoot, is widely used and understood by many. This technique is employed in the vast majority of remote control devices that accompany televisions, stereos, and other home appliances. The infrared transmitting device must be pointed directly at the receiving device to form a communications path. A button is pushed and held down to initiate and continue a flow of power which generates the infrared beam. The infrared beam continues to be sent until the power flow is stopped or the transmission in complete. Many line-of-sight transmissions actually travel at a slight angle. The transmitting device, nonetheless, must be pointed directly at the receiving device.

Diffused propagation sends light waves to receiving devices located within the same room as the transmitting device. However, the communications path need not be line-of-sight. This means that the transmitting device does not have to be pointed directly towards the receiving device. Diffused propagation bounces light waves off reflective surfaces and onto a receiver which is connected to a personal computer, printer, or other receiving device. Since light waves cannot penetrate walls, ceilings, or floors, they remain within a room and bounce off these surfaces until they reach their destination point or lose energy.

Infrared transceivers can send mobile data at speeds up to 16 Mb per second. They are usually packages with a converter that transforms electric current into light waves and, conversely, light waves into electric current. Light waves combine to form a light beam. The power generated to form the beam is a major determinant of the volume of mobile data the beam can carry and how far it can travel. Since infrared communications are known to interfere with some other electrical equipment in hospitals and factories, most products use low power transmission. High-powered transmissions are generally limited to line of sight.

Infrared Products for Wireless Computing

Infrared technology is generally used in three types of products: mobile communicators, laser diodes, and local-area networks. Mobile communicators are small, portable, low-power devices used to send information while moving about. In the world of wireless computing, mobile communicators are the equipment installed in computing equipment or packaged in converters that support the Infrared Data Association's (IrDA) protocols (see Chapter 6—Portable Hardware – Peripherals and Integration). IrDA has developed three protocols for infrared communications—the Serial Infrared (SIR) protocol, the Infrared Link Access Protocol (IrLAP) and the Link Management Protocol (IrLMP). The IrDA protocols are based on point-to-point communications and transmit at a rate of 9.6 kbps to slightly less than 5 Mbs. The IrDA-based mobile communicators enable mobile data to be exchanged wirelessly between personal computers and printers, personal digital assistants and automatic teller machines, or palmtop and desktop computers.

Laser diode communications require lots of power, transmit over line-of-sight, and are generally used by the military as a alternative to radio frequency or landline communications. These systems use high-power infrared communications to send larger amounts of information over longer distances. Some laser diode systems are under development to provide campus area wireless communications commercially.

A few companies have introduced local area networks that employ diffused propagation infrared technology. As is the case with some other wireless local area networks, diffused propagations LAN products have not yet established themselves. As a matter of fact, one company, Photonics, went out of business in July of 1995. If adherence to standards is to play a role in their success, many wireless local area networks have positioned themselves for growth. Most have adopted the infrared standard developed by the 802.11 committee of the

Institute of Electrical and Electronics Engineers (IEEE). These wireless LANs use diffused propagation, transmit at a rate of 1Mbs and have a longer range than the lower powered mobile communicators.

Infrared communications are used in transmitting mobile data for local-area networks and short range transmissions (e.g., PC to printer). While, technically, it is possible to send information via infrared transceivers that are physically connected, line-of-sight and diffused propagation characteristics of infrared technology do not make wide-area communications technologically feasible. The disparate nature of infrared characteristics requires a separate standard for each. As a mature form of wireless communications, with only two standards that are endorsed by a large number of vendors, infrared will remain a technology of choice for selected functions and products.

Communications Software

Providing Functionality for Wireless Communications

Wireless communications software requires a different set of functions than software designed for landline communications. It will be helpful to highlight key differences between wireless and landline communications.

A communications network that employs cables, wires, and other ground-based physical devices to transmit data is considered a landline network. For many years, the computer industry has used landline networks to transmit data. The physical network components included computers, modems, PBXs, repeaters, boosters, multiplexers, or concentrators. These components provided the power and much of the functionality required to move data through the wires and cables that connected them. These components, by nature, are located below, slightly above or on the ground's surface. When assembled together, they form a line which travels along the land's surface. Thus the name, landline communications.

Landline communications provide a pipe or channel (using wires, cables) through which a fairly constant flow of data can be maintained. This pipe has defined physical boundaries to contain the flow of information. These boundaries protect the data flow from outside elements (i.e. wind, rain, gravity). Most outside elements cannot penetrate the boundaries. Physical components within the boundaries (i.e. repeaters, boosters) may be used to assist in maintaining the speed of the data flow. Large volumes of data can be sent through landline networks at high speeds and at a low cost. The process of transmitting data via landline communication networks is widely used and understood by many.

Wireless communications, on the other hand, uses electric current and radio waves to send information through the air from one point to another. As we learned earlier, information is usually broken up into packets and sent through the air, across selected frequencies, until all packets are received at the destination point. Unlike landline communications, wireless requires a conduit without boundaries or protection—that is, air. When mobile data travels through the air it may encounter obstacles which can cause damage or even obstruct its voyage. Wind, rain, and sun can damage mobile data transmissions. Buildings, trees, and hills can halt transmission. Moving the portable computing device can affect the transmission of mobile data. Many factors which may be unknown to the mobile consumer can impair wireless transmissions and cause the signal to disappear without warning. The radio line could be lost seconds after making the connection. If this occurs, the connection must be reestablished so that the mobile data can be retransmitted. The need to manage pieces of information traveling through an uncontrolled, ever-changing environment takes on an added dimension of complexity. Development challenges include changing latencies, handling flow control, incongruous coverage, and interrupted transmissions, among other things. This complexity is of a greater magnitude because wireless communications does not enjoy the luxury of plentiful space, or bandwidth, as do landline communications.

When used properly, communications software can play a major role in controlling the cost of wireless communications. When wireless or landline local area networks are used, you pay the cost of communicating when you purchase the network. After the network has been installed, there is no additional charge for the transmittal of mobile data over a local area network.

With the exception of support and upgrades, wireless and landline local area networks provide substantial bandwidth (i.e., 10 Mbps) at a fixed price. Wide-area public networks, on the other hand, charge for what you use. Whether you use a circuit switched network (i.e., distance or time-based pricing) or a packet switched network (i.e., volume-based pricing), you will pay usage fees each time you use the network. Circuit switched networks charge for bandwidth and packet switched networks charge for packets. Although these networks offer a moderate amount of bandwidth (i.e., up to 19.2 kbps for packet switched), the advantage is that you only pay for what you use. Communications software allows the mobile consumer to make the most of this advantage by ensuring that the minimal amount of mobile data is transmitted. Thus, you only send and pay for what is absolutely needed.

Enabling Wireless: Middleware and Beyond

Communications software which enables the transmission of mobile data over wireless networks is commonly referred to as wireless *middleware*. This group of software can range from individual software packages, which perform selected communications functions, to groups of applications programming interfaces (APIs), which can be embedded in software programs. Whether it allows you to send mobile data from the **Print** command in Windows, or from within an application program, wireless middleware is essential to wireless computing. Wireless middleware provides the applications software developer with the tools to enable communications cost-effectively over wireless networks. These tools greatly facilitate the development of applications software which offer wireless communications. Wireless middleware is necessary for the transmission of mobile data over wireless networks.

The efficiency of communications software plays a key role in the speed and accuracy of the end-to-end transmission of mobile data over wireless networks. The manner in which communications software works with applications software, coupled with its proficiency in sending mobile data over wireless networks, determines its level of efficiency. It also helps to minimize the costs of wireless communications. Applications software should be able to easily identify and transfer data to communications software. It should also be able to give communications software prioritization and transmission instructions without intervention from the mobile consumer or the host consumer (i.e., user). When communications software works with applications software, the operating system, and the wireless network, it can play a major role in facilitating the efficiency of wireless computing.

As with all software, communications software should strike a balance between ease of use and functionality. To maximize ease of use, a simple graphical interface can be provided for mobile consumers or programmers. The person operating the computer would identify the files or data elements that are to be transmitted wirelessly and provide instructions for sending the mobile data (which network, what time to send, and so on). Most E-mail software functions in this manner. A simple graphical interface is often preferred by individual consumers who are not using wireless computing for time-critical applications. Delivering this level of functionality to the mobile consumer would, however, increase the steps to operate the total application and increase the overhead associated with performing wireless communication functions. Although minimal, additional time, training, and computer resources would be required to execute the mobile application. Even minimal time is of the essence for most wireless computing applications.

To maximize functionality, the wireless middleware must deliver functionality to programmers that allows them to embed communications functions in the applications software. These functions are delivered in the form of application programming interfaces (APIs). The overhead necessary to launch a graphical interface and perform transmission instructions is eliminated. The programmer can invoke only those communication routines which are required at a given point in the application. The interface and structure of APIs must be designed to minimize the amount of time that a programmer must spend to understand and use them. This must be achieved without substantially impacting functionality.

The art in striking the proper balance between ease-of-use and functionality is to provide a product which best meets the need of the target market. Whether the product is intended for Independent Software Vendors and application developers or the mobile consumer, balancing functionality and ease-of-use will help to determine the success of wireless middleware.

In providing for the transmission of mobile data over wireless networks, communications software should strive to achieve several goals, including but not limited to:

- Maximizing data compression
- Applying client server techniques
- Using queued messaging paradigms
- Processing as much as possible before transmittal
- Avoiding terminal emulation

Communications software performs several functions that are integral to the transmission of mobile data over wireless networks. These functions include transmission management, gateway management, application routing, and network optimization.

Transmission management includes initialization and monitoring of the radio modem, queuing data messages or packets, data compression, and ensuring the transfer and receipt of all data messages or packets. Performing these functions is much more complex than handling functions for landline communications for several reasons. Software functions for wireless communications must manage random activities as opposed to the predictable and sequential nature of landline communication activities. Additionally, many of these functions must be performed simultaneously. Transmission management processes provide the foundation for sending data over wireless networks. They enable the accurate, complete, and efficient transmission of mobile data.

Gateway management includes protocol conversion, linking mobile data to host applications, and interfacing with host-based radio modems and communication paths. This should be achieved while coordinating messages at the host from multiple mobile consumers. These processes ensure that mobile data arrives at the host computer in a usable format.

In order for multiple applications to coexist and communicate over wireless networks, the underlying software must know how to handle *packet routing*. This allows the mobile consumer to communicate with several parties simultaneously. If you have data coming in from different sources for several applications, the communications software should allow the data to be routed to the right application. Applications routing empowers the mobile application to designate which host applications should receive the mobile data. It can also designate which networks should be used for the transmission of mobile data. In applications routing, the selection of the network is dependent upon the host application's ability to access it. That is, the host computer must have an active gateway to the wireless network which is delivering the mobile data. These processes allow the mobile data to arrive at the appropriate location(s) while minimizing transmission costs.

Network optimization allows the mobile and host applications software to provide parameters for selecting which wireless networks should be used for the transmission of mobile data. These parameters may include the volume of mobile data, network transmission costs and network availability. Ideally, communications software should reduce the resource consumption and complexity of performing these processes by taking control of the communications ports on the host computer and the mobile computing device. These processes increase the timeliness of receiving mobile data and also reduce transmission costs. While the functions of transmission and gateway management, applications routing and network optimization together provide a robust foundation for wireless middleware, no single communications software package provides all of these functions for all wireless networks. A few vendors are delivering wireless middleware, however, and future development is likely to result in software products which deliver these functions in a cohesive or interoperable group of software packages.

Although the existence of communications software is essential to the growth of wireless computing, its development has been hampered by several challenges. A well-defined set of standards has not yet been established by the Computing Communications industry. Herculean efforts are underway, however, by industry associations and vendors to embrace a group of functionally efficient protocols. The proliferation of existing and emerging networks requires that communications software vendors develop several links and transmission routines. Communications software must compensate for the lack of mobility functions in existing operating systems or adapt to the mobility functions of new operating systems. And finally, the limited availability and development of applications software reduces the revenue stream which provides the funds and justifies additional funding of development.

Since network carriers and hardware vendors have the largest investment in wireless computing, it is in their interest to subsidize the funding of development of communications and applications software. As the demand for wireless computing systems continues to grow, there will inevitably be a growth in the number of communications software vendors as well as in the functionality of the software they provide. When the market begins to mature and stabilize, shrink-wrapped versions of this software are likely to emerge. In the mean time, the short list of vendors who deliver wireless middleware will reap the benefits of operating in a market segment with limited competition.

Wireless E-Mail Software Packages

Electronic mail, or E-mail, software has grown in popularity as increasing numbers of consumers communicate over computer networks. E-mail software is used on mainframe or mini computer-based networks, local-area networks, and value added network (VAN) services. As the number of subscribers and their demands grow, network providers have increased the number of gateways and services (e.g. access to the Internet and databases, facsimile service) that are available. The functionality of E-mail software has also increased. E-mail has changed the fabric of computing communications with its growing popularity and increase in services.

The same phenomenon is taking place in wireless computing. As the number of mobile computing consumers grows and their demand for wireless communications increase, we have seen the emergence of wireless E-mail software and services. A couple of wireless E-mail software packages, which also support local area networks, have hit the market. These are among the first shrink-wrapped software packages that support communications over wireless networks. In addition to the standard E-mail functions, which include sending and receiving messages, attaching files, and broadcasting messages, these packages also offer functions that enhance their performance over wireless networks. Rules-based messaging, data compression, and message monitoring are a few of the features which have been enhanced or added to improve the performance of wireless messaging. Additional features for wireless E-mail software include interfaces to local-area networks, message preview and message filtering.

In addition to shrink-wrapped software packages, wireless E-mail software is also delivered in conjunction with wireless messaging services. These value-added networks offer several services that are common to all, including store and forward messaging, an Internet E-mail address, paging access, wireless facsimile and message center operators. Additionally, selected providers offer unique services such as text-to-speech conversion, network independence, and automatically forwarding messages on the fly. The E-mail software that is bundled with the services of these value-added networks is quite functional, has a friendly interface and is easy to use. In most cases, the E-mail software was developed by the same company who is independent of the value added network provider. While this segment of wireless computing shows much potential, the entry costs are not insignificant and only three companies offer these services.

Operating Systems Software

Functionality

The *operating system* provides the intelligence to direct the computer how and when to use its resources. It is the resource manager; tells the computer's processor (chip) what set of instructions to execute. These instructions include interpretating information input from the keyboard, touch screen, bar code wand, or modem; activating memory to provide additional space for processing instructions; or sending data to a storage location or to a modem for transmission to another party. Since the operating system consists of computer code or software, it requires a storage location in the computer. In portable computing devices, operating systems are usually stored on a hard drive or a computer chip. The operating system can also reside on a PC card.

The power of an operating system can be measured by the usefulness of the functions that it can perform and how efficiently it performs these functions. A classic dilemma faced in the design of operating software is to strike the right balance between resource consumption and functionality. This dilemma is of particular concern when designing a software interface for an operating system developed for portable computing devices.

The software interface should provide functions which greatly facilitate use of the portable computing device by the mobile consumer. It should offer functional simplicity. The goal of the designer is to deliver robust functions which are intuitively easy-to-understand and operationally easy-to-use. This goal is impossible to achieve without consuming valuable resources from the portable computing device (e.g., memory, battery power, storage). Herein lies the dilemma. To what extent does the design of the software interface incorporate functional simplicity and still provide a foundation for applications software to perform within the limited resources of portable computing devices? Indeed, there is no simple answer to this question. As you will learn later in this Chapter, some companies have responded by using object-oriented programming technology to design and develop their operating systems.

Two early measures for predicting the commercial success of an operating system are:

- The number of computing devices that support it
- The influence of the independent Software Vendors who port their software to it

Operating system vendors bring strength to these measures through several efforts, including effective marketing, substantial financial resources, and a quality product that delivers marketable features and power.

As a set of software programs, the operating system uses memory, storage, and processing power to perform. One of its primary objectives is to provide as much functionality as possible to applications while minimizing the amount of resources consumed in doing so. The richness of an operating system is determined by the set of features that it delivers while maintaining this balance.

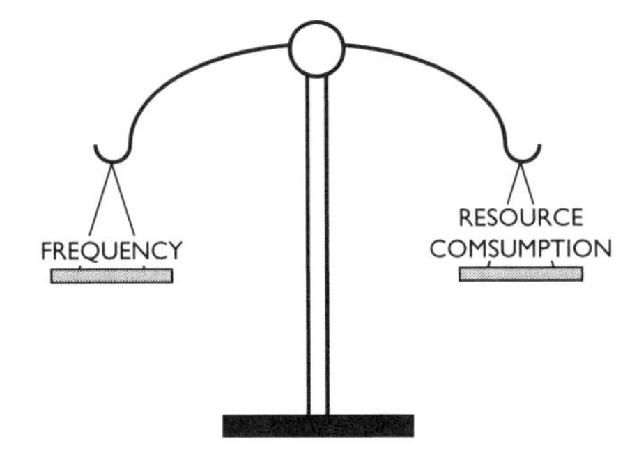

Figure 4.1 Operating System Balancing Act

Some operating systems take a *modular*, or layered, approach to delivering functionality (see Figure 4.1). A basic set of functions is embodied in its core, also known as the *kernel*. More advanced functions are available in subsequent layers, or realms, of the operating system. This allows the applications designer or systems manager to minimize the amount of storage consumed by only loading those modules or layers that are necessary to run selected applications. UNIX and MS-DOS can be apportioned in this manner.

Other operating systems use an object-oriented approach. Although the entire operating system is loaded on the computer, the amount of storage that it consumes is often moderate in relation to its functionality. This is because the objects, which comprise the operating system, perform a variety of functions. Since these objects can be moved in and out of memory as they are needed, object-oriented operating systems generally use RAM and ROM more efficiently than operating systems which do not employ object-oriented programming techniques. The Newton operating system, Magic Cap, and GEOS are examples of object-oriented operating systems.

The functions that an operating system provides can be grouped into four categories:

- Deliver a user interface
- Manage resources
- Communicate with applications software
- Communicate with external environments

Deliver a User Interface

The *user interface* is the packaging of an operating system. It is the defining characteristic that allows the end user or mobile consumer to visually distinguish it from the competition. The interface helps the mobile consumer understand what functions are available from a computing system. Since graphical user interfaces (GUIs) simplify and expedite the comprehension process, they have become a popular component in personal computing systems. To be of value, an interface must be easy to understand, programmable, and fast. Operating systems should deliver comprehensible and efficient interfaces.

Manage Resources

Resource management is the operating system's *raison d'etre*. The operating systems functions as a traffic cop to determine when and how the resources of the computer will be used. The computer's resources include memory, storage, screen display, and peripheral devices. The operating system delivers functions that protect the applications software from having to perform the rudimentary task of communicating with the hardware. This includes invoking memory and managing its use (i.e., swapping) to get the most out of what is available. The operating system must tell the computer processor or chip when it will receive data and how it should act on it. It must track peripheral devices and manage the flow of data to and from these devices. Whenever possible, the operating system should allocate resources (i.e., power) needed by the primary computer and its connected peripheral devices. To effectively manage the hardware components of a computing system, the operating system should be able to adapt new technologies, manage all peripheral equipment, and use functional processes as they are needed. It must also resolve conflicts between the computer resources in a manner that optimizes the computer system's performance. As the resource manager, the operating system orchestrates the processing and movement of data within the computing system.

Communicate with Applications Software

The operating system provides applications software with the tools it needs to function. It provides the language translation necessary to allow the applications software to communicate with the assembly language of the computer's processor or chip. To facilitate this process, the operating system should notify the applications software when certain functions have been completed and when certain situations arise (i.e., battery level is low). The operating system should also provide a means (i.e., OLE or object linking embeding) for different applications software packages to communicate with each other. In short, it should provide a core set of functions that assist Independent Software Vendors (ISVs) and application developers to enhance the performance of their software.

Communicate with External Environments

Computers are no longer individual machines that operate in their own domain. In order for the information produced by computers to be of value, it must be communicated to the outside world. To this end, individual computers are often merely components in enterprise computing systems. At the very least, the information produced by an individual computer should be provided to others. In order to make the information or data from a computer available to external environments (i.e., enterprise computing systems, or individuals), communications functions must take place. These functions include network communications, file transfer and synchronization, and protocol conversion.

An operating system generally facilitates network communications by efficiently managing the physical devices (i.e., modems and serial ports) that connect to networks. To minimize learning and maximize ease of use, the process of communicating with these devices should be accessible from the user interface (i.e., MAPI). File transfer and synchronization should be managed by the operating system via the user interface. By managing protocol conversion, the operating system converts data into a format that can be read by another person or computer. To facilitate communications with external environments, the operating system should manage the devices (i.e., modems) and rudimentary functions that are necessary to send data in a form that can be understood.

Facilitating Mobility

As a resource manager, the operating system should coordinate the use of hardware and perform the basic functions necessary to handle the processing and movement of data. In a wireless or mobile computing system, these functions take on an added dimension. The challenge of the operating system on the mobile end is maximizing its functionality and power within the physical confines and limitations of the portable computing device.

Operating systems that support portable computing devices must also incorporate "mobility" functions to increase their acceptance by mobile consumers. These functions are essential to provide mobile consumers with the processing and communications they require while containing the effort necessary to use the mobile system.

Mobility functions include:

- Power management
- Peripheral management

- Advanced communications
- Desktop coordination

Power Management

Since battery life is a precious commodity in portable computing devices, the efficiency of power management is important. The operating system should monitor the battery level and minimize its consumption by hardware components. Early palmtop computers allowed the batteries to be changed without turning off or powering down the computer. Many portable computers now incorporate a variation of this functionality.

Power management is most efficient when the applications software operating system, and processor work in concert with each other. Advanced power management (APM), for example, is a tool that helps developers write power-efficient applications software. APM is the result of a collaboration between Intel and Microsoft and is available for Windows. An operating system can manage hardware and applications software in a manner that maximizes the power in all portable computers.

Peripheral Management

Early palmtop computers allowed PC cards to be inserted and removed without turning off the machine (i.e., *hot swapping*). Likewise, the docking stations for some early mobile computing systems were designed to allow data to be transferred and batteries recharged when the portable computing device was inserted (i.e., *hot docking*).

Advanced Communications

In addition to being able to perform communications while working on another application, many computing consumers have become accustomed to using multiple modes of communication. Not only do mobile consumers want this same functionality, they also need to communicate over wireless networks. Some hardware vendors have integrated radio modems into their portable computing devices, and many support PC card radio modems (i.e., PCMCIA Type III). To facilitate the transmission of mobile data over multiple networks, operating systems should provide functions that enable the dynamic redirection of mobile data. This would allow the mobile consumer to set criteria that governs when and how mobile data is to be sent. The operating systems should provide the base functions necessary to enable communications software to manage the transmission and cost of sending mobile data over wireless networks.

Desktop Coordination

Some database and contact management software supports the transfer and synchronization of data elements and files between desktop and portable computers. The operating system should provide the underlying functionality to facilitate this process.

Mobilecentric operating systems are required to further promote the use of wireless and mobile computing. Operating systems have emerged that support many of the mobility functions that wireless and mobile computing systems require. In addition to providing these functions, several do so with a level of efficiency that allows them to perform within the platforms of smaller portable computing devices.

Windows 95

Windows 95 has a very robust environment for mobile consumers. It supports peripheral, power, and communications management functions including hot docking, PC card dynamic plug and play, deferred printing, battery monitoring, and dial-up networking. Its messaging (MAPI) and telephony (TAPI) applications programming interfaces contribute to improving the ease and efficiency of communications. Windows 95 is designed to run on notebook and some palmtop computers.

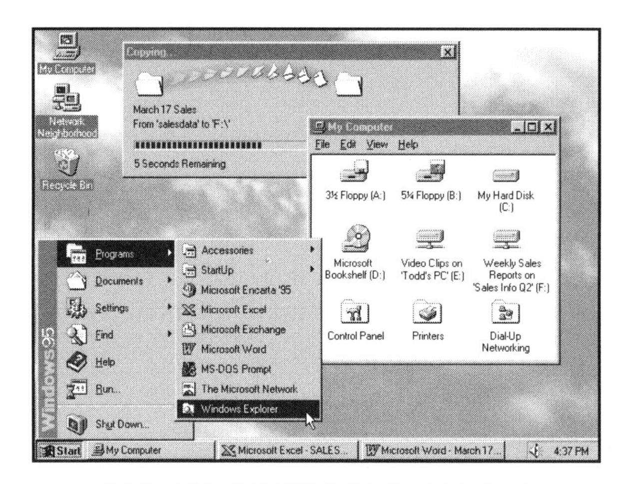

Figure 4.2 Windows '95 Interface.

GEOS

GEOS provides a highly functional communications environment for personal digital assistants and computer tablets. Its socket-based and session-based features are analogous to packet-switched and circuit-switched network technologies, respectively. Socket-based features enhance the applications processing environment by allowing several applications to share the same socket or communications path. GEOS also provides an OUTBOX for self-contained application messages (i.e., message-based) and file transfer capabilities.

COMPONENT LAYERS OF GEOS	
APPLICATIONS	GEOWRITE™,GEOFILE®, GEOCALC™
SYSTEM LIBRARY	UI, GRAPHIC, SPREADSHEET OBJECTS
KERNEL	SCHEDULING, MEMORY MANAGEMENT
DEVICE DRIVERS	VIDEO, MOUSE, PRINTER, FILE SYSTEM
HARDWARE	SERIAL/PARALLEL PORTS, PCMCIA,EMS/XMS

Figure 4.3 GEOS component layers.

Magic Cap

Magic Cap has a fully functional and intuitive user interface that allows the mobile consumer access by touching the screen with a finger or stylus (see Chapter 6, "Portable Hardware"). It is designed to exploit the functionality of personal communicators by offering E-mail, fax, telephone, paging, and infrared support. As with GEOS, Magic Cap's object-oriented design allows applications to maximize their performance in the resource-constricted environment of personal digital assistants. Exceptional memory management facilitates high-performance functions in a multitasking environment.

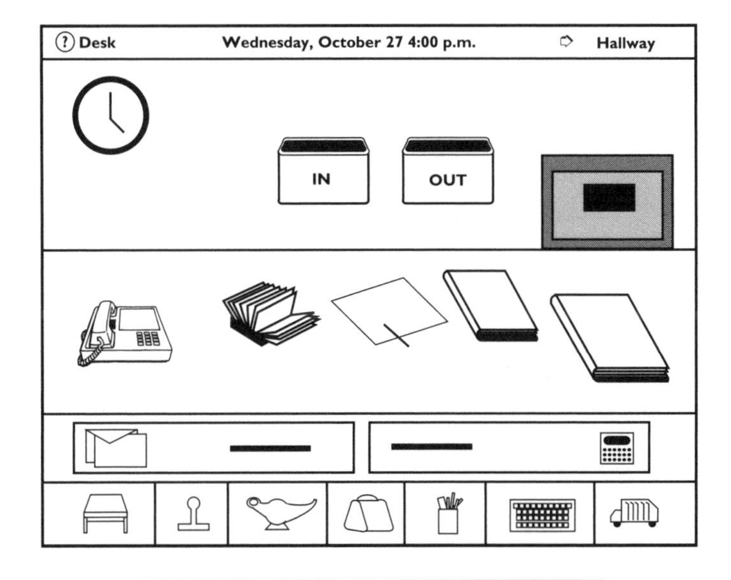

Figure 4.4 Magic Cap user interface.

Newton

The Newton operating system uses an object-oriented design to provide functionality which maximizes the performance of personal digital assistants. It includes a micro-kernel, a memory management subsystem, and support for removable storage and input output (I/O) devices. The micro-kernel is structured to deliver core functions for applications software while minimizing the use of computer resources. Its memory management unit (MMU) delivers a high level of performance by using small blocks (i.e, 1K), automatically freeing unneeded space and allocating storage as it is needed. In addition to enabling the support of wireless communications, the Newton operating system dynamically recognizes peripheral devices such as PC cards and those connected via a serial cable.

STRUCTURAL OVERVIEW

Packages	
Object-based Storage	
Heap and Stack Management	Read-only device Management
Memory Architecture	
Micro-Kernel	

MICRO-KERNEL OVERVIEW

Monitors	Semaphones
Tasks	Ports and Messages
Tasks Scheduler	Timing Services
Object Manager	

Figure 4.5 Newton O/S Structural Overview

Applications Software

Designing for the Mobile Environment

Mobility Features

Applications software is the essence of every computing system. It provides the core functionality that gives the system its central value for the consumers who ultimately use it. Applications software provides the main definition of exactly what the mobile consumer can achieve with their mobile or wireless computing system. As with stationary computer systems, the applications software's ability to interface with systems software (i.e., operating systems and communications) is a major determinant in the performance of the system.

Most popular personal computer software packages designed for desktop computers can also run on notebook computers. Whether it runs on DOS, WINDOWS or WINDOWS 95, these software packages can perform on notebook computers. Many notebook computers offer the performance and resources (e.g., memory, processor, storage) of typical desktop PCs. With the exception of notebook computers, most portable computer systems have limited resources (i.e., memory, storage, peripheral ports). The operating system manages these resources. To enhance performance, it is necessary that the applications software work hand in hand with the operating system to maximize the use of the limited resources of portable computing devices. Applications software must use memory management routines that are available from operating or other systems software. The effective use of memory is essential to attain a response time that will satisfy the consumer on the move. Storage in portable devices is usually available from PC cards or hard disk drives with capacities are far smaller than those found on desktop computers. Storage is used to house applications software and files which that software creates or uses. Some applications software is delivered on PC cards as well as on diskettes.

Whether it is located on a PC card or on the hard drive, the applications software should use operating system functions to recognize the location and capacity of storage available. To the extent possible, applications software should notify the mobile consumer when a PC card or other form of static storage should or should not be inserted. Mobile consumers are increasingly carrying more peripheral devices to expand the functionality of their mobile systems. These peripheral devices share a limited number of ports. If the device drivers, operating systems, and communications software permit it, the applications software should provide screen messages which indicate what peripherals are expected to be used on which ports. It can also determine which mobile data should be sent as well as when and how it should be transmitted. For example, data produced by the application may have different time values. The applications software can be programmed to determine what data should be sent wirelessly and what should be sent over landline networks.

When portable devices do not use a standard keyboard (e.g. QWERTY), the applications software must be written or ported to recognize a different form of input. This input could be provided by a pen stylus, a hybrid keyboard, buttons on handheld devices, bar code wands or scanners, touch memory, or an infrared port. The applications software must recognize the source and format of the data input from each sources. It must also know how to use the data.

The challenge that is faced by applications software running on a mobile computing platform is maximizing functionality while minimizing the resources required to produce that functionality. That is, the software must be able to perform in a portable platform with limited processing power, memory, and battery power.

The applications software that you select is critical to the success of a wireless computing system. This is the component that drives the value of the mobile data. It is also a major determinant of the ease of use of the device carried by the mobile consumer. A major challenge in selecting the applications software is balancing the functionality needed with the usability of the portable platform on which it resides.

Defining Data Elements

To select the proper software, you must begin with a definition of the information which must be collected from or provided to the mobile consumer. The definition should include:

- The name of data element
- The size (in bytes) of the data element
- The source of the data element
- Frequency of update

Figure 5.1 shows an example of a form that you may want to create to handle such evaluation.

Mobile Data Definition Chart

Data Element	Size[1]	Frequency of Availability	Time Value	Frequency of Update	Source[2]	Delivered From Whom	Delivered To Whom

1 Bytes
2 Protocol and Format

Figure 5.1 Mobile data definition chart.

The data elements required depend on the business requirements to be achieved. Whether you are collecting package tracking numbers and customer signatures or receiving current stock prices and trading volumes, every single data element needed to satisfy the business requirement must be identified. As the data elements are being identified, several characteristics of each element should be determined. These characteristics are major determinants of the speed and cost of mobile data. It must be noted that it is usually not necessary to transmit all mobile data elements via wireless networks. As a matter of fact, the most efficient and successful wireless computing systems only transmit mobile data that is time essential. That is, those mobile elements which have a high time-value of information.

As each data element is being identified, it should be determined how large it is, where it comes from, and how often it must be updated. The size of the data element may vary from transaction to transaction. If the size is not constant, measure the minimum, most likely and maximum size. This will allow you to determine how the applications software should process the data elements.

The source of the data element is also important. The source will allow you to identify the protocol and format of the data element. It will also assist in determining who the data element was received from and how it will be used by the applications software. These elements may be "physically" input into the mobile device from a keyboard, pen, button, or bar code wand. The data could be received from a wireless transmission or via a connection to an infrared port. Mobile data elements could be input into the portable device from other internal or peripheral sources such as touch memory, landline communications from a serial or RJ11 port, a PC card, or a portable tape unit. Knowing the source of the data element will allow the applications (or operating system's) software to recognize it and understand how to process it.

The time-value of the data element will determine how often updates should be processed. The data definition should identify how often updates are *available*. The applications software can determine when and where to poll for data updated. The applications software can be instructed to poll for data if prespecified situations occur or certain data elements have been received by the mobile device. For example, if a repair technician has reached the customer's location and a credit card is the preferred method of payment, the software may automatically transmit a request to a credit service bureau over the wireless network for credit approval.

Shrink-Wrapped for Mobility

Vertical: Application Specific

As the demand for mobile and wireless applications has increased, so has the availability of software which services these applications. Several companies have developed software which addresses the needs of specific industries. Service and dispatch software providers have released versions which run on portable devices. Contact management and sales-force automation software has also been developed for portable platforms. Several software packages have been designed to meet the needs of specific industries or applications including field service, pharmaceutical, consumer packaged goods, and route delivery. Many vendors have wirelessly enabled their mobile software. That is, they have included the functionality necessary to send data over wireless communications networks.

Much of this software has matured to the point that it requires little or no customization. The vendors who provide these products offer services to adapt this software to meet the unique needs of each company. These services are provided directly from the independent software vendor or from resellers. Although the cost and amount of customization is substantially less than developing new software, this software is not shrink-wrapped and, therefore, is not available through major distribution channels (i.e., computer stores, distributors, mail order). As the wireless computing market develops, popular protocols will establish themselves and become stable. Confidence in hardware platforms and wireless technologies will grow as consumers demonstrate their preferences with their purchasing power. Independent software vendors must spend considerable time and resources to deliver a product to market. This is even true for modified versions of existing products. In addition to programming, ISVs must also lay out upgrade plans, establish support, prepare documentation, design packaging, make distribution arrangements, and prepare the product launch, among other things. The investment in any new software product is substantial. Once software vendors become confident that their investment will yield a respectable return, shrink-wrapped software for vertical business applications will begin to appear. Until that time, however, several products do exist which provide the functionality required to implement wireless computing systems with measurable results.

Horizontal: General Purpose

Horizontal, or *general purpose*, software packages are designed to meet the needs of large quantities of consumers who have varying application needs. They often provide a specific function such as contact management, remote data transfer, or E-mail. The growing interest and demand for wireless computing has moved a few ISVs to introduce products which include wireless communications capabilities.

Wireless E-mail packages have received the most attention. Some of this software is bundled with subscriptions to wireless services. These packages run on popular computing platforms (e.g., MS-DOS or Windows) and, therefore, can be easily installed in notebook computers. With a radio modem that is configured for the wireless network that these services use, the mobile consumer has virtually nonrestricted access to E-mail services. Well-established, shrink-wrapped E-mail packages have offered remote access versions for several years. These versions provide dial-in access to host systems via landline networks (i.e., Public Switched Telephone Network PSTN). Using a standard modem, operating with the **AT** command set, the mobile consumer can gain access to the central mailbox through a regular telephone. Wireless communications software (i.e., APIs) can be integrated with remote E-mail packages to provide access to wireless networks. Lotus Development Corporation has added wireless functionality to its cc:Mail Mobile™ product. CE Software has introduced a wireless version of its QuickMail™ product, MobileVision, for the Marco® Wireless Communicator. More wireless E-mail software packages will be introduced as wireless computing usage continues to grow.

Several ISVs are vying for the leadership position in providing tools to help consumers organize themselves as a flurry of contact managers and personal information managers (PIMs) hit the market. In addition to appointment scheduling and maintaining customer information, these products often incorporate document management and transmission (e.g., fax and E-mail) functions. Wireless messaging is a logical next-step enhancement for these products. Mobile Office Solutions has introduced wireless messaging for the GoldMine™ contact management software package. Its GoldMine Wireless Solution is a two-way wireless remote synchronization utility that works with GoldMine for Windows. Since contact managers and PIMs are a popular product among mobile consumers, the addition of wireless messaging functions is a logical step for these software packages.

Other shrink-wrapped software packages which incorporate wireless functions are beginning to emerge. Current packages allow consumers to send paging messages from their computers and transmit documents using the **Print** command in Windows. As the market for wireless computing continues to grow and mature, more shrink-wrapped software which incorporates wireless functions will emerge.

Portable Hardware

Enabling Portability

The challenge for equipment manufacturers is maximizing the computing power of a small, lightweight, portable device. While notebook computers have advanced to the power-level of desktop PCs, stretching desktop power to most portable computing devices is beyond the realm of current technology. In order for most portable computing devices to incorporate the components (RAM, processors, graphics accelerators, etc.) used in desktop machines, their cost would be far beyond that which most are willing to pay. Furthermore, battery life would drop to unacceptable levels (i.e., 2 hours or less). The process of selecting hardware and developing software for portable computing devices differs from that for desktop and notebook computers by orders of magnitude. A completely different paradigm must be understood to be successful in this environment.

In the early years of portable computing devices, the focus was on *size*, *weight*, and *battery life*, and SWB became an acronym for measuring where a product stood among competitors. Since their introduction, portable computers have evolved from 2 hours of battery life in 17 lb. "luggable" computers to 6–100 hours of battery life in 1 lb. "palmtop" computers. As technologies which enhance SWB have become widespread, products cannot differentiate themselves with these criteria alone. Additional mobility features have emerged. Several factors must be taken into consideration when selecting a portable computing device for a wireless or mobile system.

The factors that should inform the selection of a portable computing device are its usability, adaptability, and long-term viability. Other factors include the operating system, form factor, performance features, peripheral options, and ease of integration.

Operating System

There are only a few operating systems which are supported by multiple hardware manufacturers. This is good for the consumer since it simplifies the decision process while providing viable options. It is always preferable to use a non-proprietary operating system. Although most portable hardware vendors have moved away from proprietary operating systems, some remain. These vendors should offer a compelling demonstration of the benefits to be derived by customers who purchase their systems. For certain applications they do. Other manufacturers use a hybrid version of popular operating systems. In these cases, there is also a well substantiated reason. Many handheld vendors, for example, use small screens (i.e., less than 80 x 25) and special keypads (i.e., not QWERTY). Special drivers are required since these screens and keypads are not supported by popular operating systems. The hybrid that they employ is usually an adaptation of a popular operating system (i.e., MS-DOS) which allows for a different screen and keyboard. If you decide upon a proprietary or hybrid operating system, design your application in a portable programming language (i.e., C) and, whenever possible, limit the calls to the operating system.

Form Factor

The *form factor* dictates how easy the portable hardware is to use. The shape, input mechanism, weight, and carrying options comprise the form factor. If a handheld computing device is required, the shape should easily adapt to the average human hand. In addition, the distribution of weight in the device and the location of the screen and keypad should allow it to be held easily while using it. The input mechanism is an important consideration. This determines the hardware interface.

Whether you select a keyboard, pen, button, mouse, bar wand, touch pad, or some combination as the input mechanism, it must be acceptable to the mobile consumer. Backlighting is often necessary to allow screens to be read in the poor lighting conditions often found in mobile environments. Speed, accuracy, and ease of use must be balanced against the functionality and cost of implementing and maintaining the input mechanism. As a rule of thumb, handheld devices should not weigh more than 1 pound. Portable devices, including all peripherals, should weigh less than eight pounds. Despite the functionality and benefits that a mobile computing system may afford, if it is too large and heavy to be carried easily it will not be used as often as desired by the mobile consumer.

Carrying cases, shoulder harnesses and belt holsters have become popular means of carrying portable computing devices. Hardware manufacturers have also increased the portability of their devices by adding carrying straps and handles. Some vendors have even modified the design of the casing to include grooves or impressions to make their products easier to carry. These indentations in the casing allow the mobile consumer to grasp and hold the portable computing device with ease. Their hands and fingers can better conform to the exterior of the casing. When selecting a portable computing device consider the options available that make it easier for the mobile consumer to carry.

Performance

The *performance* of a portable computing device will be determined by the power of its components as well as its ability to withstand the rigors of the mobile environment. The major components which determine performance include the processor and circuitry, random access memory (RAM), read only memory (ROM), and storage.

Most portable computing devices use 3-volt processors as compared to the 5-volt processors typically found in desktop systems. A system which is designed using low voltage components (i.e., processor and circuits) will perform better. A low-voltage hardware design, however, must be supplemented with power management to maximize battery life. The basic input output system, or bios, is the pathway for sending information and instructions in the computing device. Every time that information or instructions are sent the computer consumes power. The bios needs to work in conjunction with the operating system and applications software to maximize power management.

In addition to low voltage, the hardware design often incorporates various levels of power consumption. Since a primary objective of power management is to conserve power while maintaining response time, these modes vary based on these measures (see Figure 6.1).

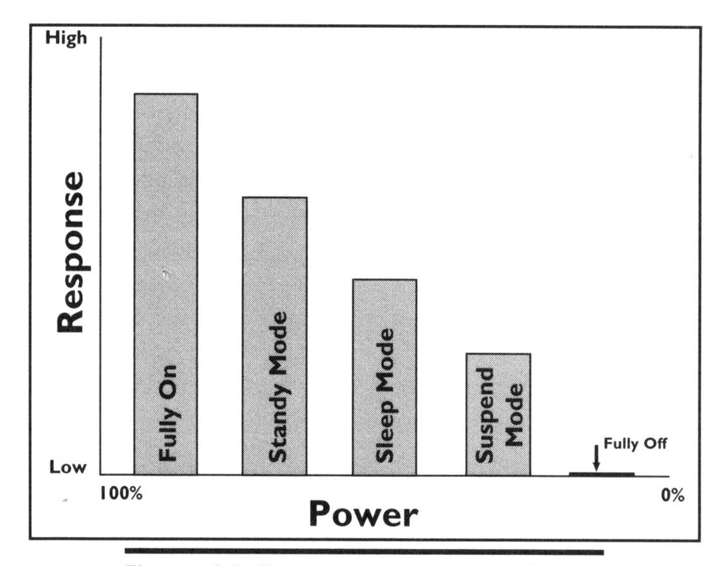

Figure 6.1 Power management levels.

The amount of random access memory, or RAM, is also a determinant of the speed at which software performs. Notebook computers often have the same amount of RAM that can be found in desktop computers. Personal digital assistants (PDAs) and other smaller computing devices have less RAM—about 1–4MB. Read only memory (ROM) is often to used to store software functions that are frequently used such as operating systems and horizontal applications software (calendar, calculator, etc.). The storage available for portable computing devices is usually provided with a hard disk drive or PC card.

The combination of processor, RAM, ROM, storage, and power management are the major factors that determine the performance, or speed, of the portable computing device.

Any item that is frequently carried by hand is subject to being dropped. Since mobile consumers are often outside, these items are also susceptible to the elements, including rain, sun, humidity, and moisture. Portable computing devices should be constructed to withstand the rigors of the mobile environment. They should be able to survive repeated drops of 3 feet or more onto concrete. Other tests of a unit's ability to withstand the environment include temperature range, humidity range, shock, vibration, and altitude range. To meet these requirements many devices are sealed, have rounded corners, and the motherboard is shock mounted. Some manufacturers construct devices which meet or are close to military specifications (i.e., MILSPEC) for mobile consumers who operate in extreme environments (e.g., construction workers). When selecting a portable computing device make sure that it will perform in the environment of the mobile consumer.

Peripherals and Integration

Peripherals extend the functionality of a portable computing device. They include PC cards (i.e., PCMCIA), modems, AC adaptors, docking stations, tape drives, and printers (Figure 6.2). Infrared ports are quickly becoming a popular feature of portable computing devices. These ports allow mobile data to be transferred wirelessly using line-of-sight infrared beams. They work like a television remote control device. You simply point the infrared port towards the receiving device (i.e. printer), and shoot. The mobile data is transmitted to the receiving device through the infrared beam.

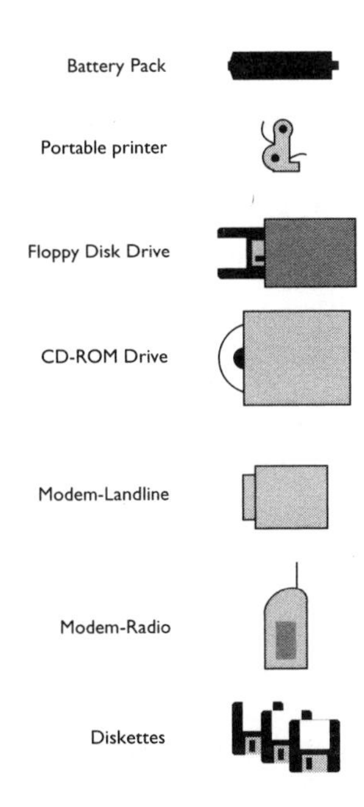

Figure 6.2 Various handheld accessories.

IrDA, the standard for infrared ports, allows transmissions of up to 10 feet. It is described bv John LaRoche "an interoperable, low cost, low power, half-duplex serial data interconnection standard". Infrared ports using the IrDA protocol enable the mobile consumer to perform several functions including file transfer to a desktop PC and transmitting documents to printers in various locations. Protocol converters, also referred to as adaptors, will allow IrDA based infrared transmissions to be converted into signals which can be received by a RS232 serial port. IrDA based adaptors, therefore, allow an infrared port to be added to existing equipment.

Portable hardware devices which support wireless computing systems must include a radio modem. Radio modems can be external, in a PC card (i.e., PCMCIA Type III) form, or fully integrated into the computing device. Regardless of the manner in which radio modems are connected, the systems designer must employ power management routines. Although most radio modems have their own battery, many also draw upon the power of the portable computing device. Radio modems can be connected in several ways. They may be attached using a cable mounted with velcro. PC card modems can be inserted into a slot in the portable device. Finally, several OEMs fully integrate radio modems into their hardware. Personal communicators (see Personal Digital Assistants) integrate radio modems into their platforms. Since radio modems are tuned to modulate and access radio waves for each wireless network, they are network dependent. That is, you must select a radio modem that works with the wireless network you wish to access.

The functionality of many peripherals (e.g., radio modems, hard-disk drives) are being delivered in PC cards. The portable computing device must be designed with the appropriate number of ports (i.e., serial and parallel) and PCMCIA card slots to handle the peripherals that are required by the mobile consumer. Keep in mind that while many peripheral devices have their own battery, these devices often draw upon the battery of the portable computing device. It is advisable to select portable computing devices with standard ports (i.e., serial, parallel, PCMCIA) since those adaptors or converters which allow standard peripherals to plug into proprietary ports are often expensive.

PCMCIA Cards

Version 1-memory cards. 5V EPROM/FLASH
Version 2-expanded to include I/O, 3V

Type:

1. **3.3mm, ROM, RAM**
   EPROMS, flash RAM
2. **5mm, adds modems, LAN, SCSI, sound**
3. **10.5mm, Hard drives, Radio Modems, etc**
4. **16mm, not yet approved**

Figure 6.3 PC Cards (PCMCIA).

For some applications, the portable computing device must be integrated with other equipment to provide a hardware platform that is functional for the mobile consumer. If this is the case the portable hardware design must incorporate features to facilitate integration. Considerations include the location of the PC card slots, peripheral ports, input and display mechanisms (i.e., keyboard and screen), and battery or power connections. Integration can often be made easier if the portable device has screw bosses, which allow other equipment to be physically attached to it. When integrating other power consuming equipment with the portable computing device, the designer should incorporate power management techniques to ensure that the mobile consumer has adequate battery life. The systems planner should also make preparations to provide support and maintenance for the newly integrated device.

Portable Computing Devices

As equipment manufacturers tackle the challenge of maximizing the functionality and power in a small, lightweight, portable device, five categories of mobile hardware have emerged:

- Notebook Computers
- Palmtop Computers
- Personal Digital Assistants
- Handheld Devices
- Tablet Devices and Pen Computers

Each type of device was developed to meet a different set of market needs; however, advancement in computer chips and power management have resulted in an overlap of their target markets.

Notebook Computers

Notebook computers are the most widely used portable device. Advancements in technology have resulted in notebook computers that provide functionality almost parallels to that of desktop computers.

Palmtop Computers

Palmtop computers have been on the market for several years and have amassed a respectable installed base, largely due to the success of the Hewlett Packard LX line. Although most palmtop computers do not provide an environment that supports all applications run on desktop and notebook computers (e.g., Windows), they do run conventional software programs in a compact, highly portable unit.

Personal Digital Assistants

Personal digital assistants (PDA), also called personal communicators, promise to usher in a new era of mobile computing. While these devices have yet to establish a sizable installed base, new models continue to be introduced. PDAs incorporate a highly intuitive interface, using pen technology and operating systems that exploit their advantages. PC card slots provide a conduit for compact storage and extended software functions. Most devices integrate advanced communications including access to wireless networks.

Handheld Devices

Handheld devices have long been used for warehouse and repair applications. As they adapt standard operating systems and peripheral ports, their position in the wireless and mobile computing market has grown. These devices often offer a higher level of ruggedness than other portable computing devices. Some handheld devices use pen technology for their input mechanism.

Tablet Devices and Pen Computers

Pen-based tablets and computing devices have sought to establish themselves in a niche: vertical markets. They usually include electromagnetic or resistive digitizers to receive input from the mobile consumer. Input can be made with a stylus, your finger, or a thin, solid object (e.g., fountain pen). Electromagnetic digitizers require power, and thus can only receive input from a stylus designed for the device. While handwriting recognition has greatly improved over the years, it is advisable that mobile consumers use "point-and click" to input information and operate pen computers.

It should be noted that digitizers, particularly electromagnetic digitizers, are somewhat fragile. This must be taken into consideration when evaluating pen based computers for use in harsh conditions. Although some pen-based devices are more rugged than others, many require special carrying packages to endure most mobile environments.

Notebook and palmtop computers, PDAs, handheld devices, and pen-based computers are generally lightweight and small enough to be carried by the mobile consumer. In some instances, it is preferable to mount these devices in a car or truck. Vehicle-mounted hardware can add an extra measure of ease for the mobile consumer. In addition to serving as a convenient means of storing the portable computing device while in transit, vehicle mounting also allows improved wireless communications and recharging. This is due to the fact that vehicle mounts are often designed to draw upon the battery of the vehicle. Wireless communications can be improved further by mounting the antenna on the top of the vehicle. Wireless computing applications which employ vehicle-mounted portable computing devices are common in fleet forces.

Batteries

Batteries are the lifeblood of portable computing hardware. They supply the fuel that propels wireless and mobile computing systems. Battery life is affected by virtually all facets of system operations. The complexity and amount of data processed has a direct relationship to battery consumption. Since wireless transmissions are produced by electric currents which are fueled by batteries in portable hardware, the amount and distance of mobile data transmitted also has a direct effect on battery life.

There are several types of batteries which may be used in portable computing devices including lithium, nickel metal hydride, nickel cadmium, and alkaline batteries. The trade-off in the selection of batteries is in cost, battery life and convenience (see Figure 6.4). Alkaline batteries can be easily purchased in corner stores and newsstands, which are readily available to the mobile consumer. Alkalines generally have the largest capacity but they are not rechargeable. Therefore, they can not support many portable computing devices which require a lot of power. Many hardware manufacturers have moved to lithium batteries due to their longer life and declining prices. In most cases, the selection of batteries is governed by the hardware that is purchased. Many mobile consumers will need or prefer an extra battery as an added measure of security. It is advisable to purchase extra batteries when purchasing the portable computing device.

	Cost per hour	Watt hour per Kilogram	Lifetime Recharge Cycles
NiCAD	$0.07	33-45	500-1000
NiMH	$1.50	50-56	500-1000
Lithium Ion	$1.50	78-115	500-1200

Figure 6.4 Battery Technology.

Regardless of the batteries used, there will likely be a need for most mobile consumers to recharge during the day. Some portable computing devices and peripherals (i.e., radio modem, notebook, etc.) have to be turned off to be recharged. This is because some AC adaptors can only perform one function at a time—power the unit or charge the battery. When these adaptors are used, therefore, you cannot recharge the computing device while it is in operation. Many mobile consumers carry multiple battery packs, or recharging units for backup power. A vehicle cigarette adaptor can sometimes be used for recharging.

Protocols

The Quest for Standards

The industry is working feverishly to produce universally accepted protocols. For our puposes a protocol is defined as any method, process or format that defines the way mobile data is handled or recognized. As is the case in "conventional" computing and communications, technology, competition, and societal politics or human nature make it virtually impossible for a single protocol to emerge as *the* standard. The functionality and behavior of different wireless technologies mandates that different protocols be employed to maximize their capabilities and performance. The same is true for software (i.e., operating systems) and hardware (i.e., peripheral ports). For many years, vendors embraced proprietary environments as a competitive edge. As wars over some protocols continue (i.e., CDMA vs. TDMA), the battles are fueled by concern for technical feasibility as well as protection of the installed base. Reaching consensus among a large group—particularly when money is involved—is a challenge that has evaded even the most skillful negotiators. Several industry groups and associations are working to establish "standard" protocols. Some have made great strides; however, few have been developed and implemented to allow true plug-and-play compatibility. Furthermore, advances in technology usually mandate the development of new protocols.

Standards are not set by technical superiority—they are market driven. A protocol becomes a standard once it has been embraced, and purchased, by a critical mass of customers. The history of our industry is littered with technically phenomenal products which never hit their mark. In virtually all cases, the lack of standards had nothing to do with it. Most people view adherence to standards as a means of protecting their investment. This may be a little short sighted. Over the course of history, as soon as a protocol has been established as a "*de facto* standard" another appears which is better and more widely accepted. The American National Standards Institute (ANSI) is an organization that is heavily involved in the review and establishment of standards. The necessity of ongoing development and integration will not go away. The first criteria in selecting a protocol, therefore, should be its level of interoperability. When systems are designed in a modular fashion, and selected based upon a reasonable return, the interoperability of protocols assists in protecting the investment in the system. The selection of software that is highly portable is essential. The modules requiring protocol changes can be changed individually. Software modifications can be made easier, networks can be changed with moderate effort, and hardware can be swapped without having to spend much time porting applications and systems software. If the benefits realized exceed the costs of these modifications, the need for standard protocols is superceded by the need for interoperable protocols.

The quest for standards should and will continue. Standards help vendors optimize the performance of their products and services. They also help speed the delivery of products to market. This quest also helps to define exactly what is necessary to design a protocol that is interoperable. It allows the functions and design of proposed protocols to be understood. This simplifies the development of interoperable protocols. When systems that deliver a high return to the business are selected and implemented, the need for interoperability exceeds the importance of establishing standards.

Up-and-Coming Protocols

As computing and communications technology continues to advance and converge, the emergence of protocols will continue. Our world is ripe with brilliant technologists who continually create a more efficient means of performing functions and processes. The value and justification of competing protocols will always make selecting a single one virtually impossible. It is unlikely that a set of purely standard protocols will emerge soon, if ever. We are close upon a group of "popular" closely defined protocols, however, which will approximate the benefits that the market demands from standards.

Since protocols exist in all facets of technology, and wireless computing systems are comprised of three major components—software, hardware and networks—there will continue to be a list of protocols to choose from. A list of popular protocols that are being touted for the major components is presented in the chart on the next page. Interoperability is likely to be more prevalent in popular protocols than those which have yet to gain a substantial following. The Popular Protocols Chart (Table 7.1) may, therefore, serve as a guideline for the selection of products and services which include them. This chart does not provide a comprehensive list of protocols. Rather, it is intended to give the reader a view of the scope of protocols that are available for wireless computing systems.

Table 7.1 Popular Protocols Chart

Acronym	Full Name	Used For[1]
OLE	Object Linking and Embeding	SW
WinSOCK	Windows Sockets	SW
MHS	Message Handling Services	SW
SQL	Structured Query Language	SW
ODI		SW
SPX		SW
NDIS	Network Driver Interface Definition	SW
TCP/IP	Transmission Control Protocol/Internet Protocol	SW/NW
IPX	Inter Protocol Exchange	SW
NetBEUI	Network communications protocol	SW
TAPI	Telephony Applications Programming Interface	SW
MAPI	Messaging Applications Programming Interface	SW
EMAPI	Enhanced Messaging Applications Programming Interface	SW

Table 7.1 Popular Protocols Chart (continued)

Acronym	Full Name	Used For[1]
VDD	Virtual Device Driver	SW
SLIP	Serial Line Internet Protocol	
SNMP	Simple Network Management Protocol	SW
SCSI	Serial Control Standard Interface	HW
Ethernet	Networking communications foundation	HW
RS 232	Hardware/firmware definition for serial port transmissions	HW
Parallel	Format for hardware connection, commonly used with printers	HW
PPP	Point-to-Point Protocol	
IrDA SIR	Infrared Data Association Serial Infrared Link	HW
IrLAP	Infrared Link Access Protocol	HW
IrLMP	Infrared Link Management Protocol	HW
FDDI	Fiber Distributed Data Interface (uses Fast Ethernet Peripheral Component Interconnect network interface cards)	
SNMP	Simple Network Management Protocol	NW
ATM	Asynchronous Transfer Mode	NW
Mobitex	Method of transmitting wireless data over packet networks	NW
DataTac	DataTac 4000, 5000, 6000 (Europe only)	NW
RDLAP	Method of transmitting wireless data over packet networks	NW
X.25	High speed method for communicating between network sites	NW
NAMPS	New Advanced Mobile Phone Service	NW
AMPS	Advanced Mobile Phone Service	NW
CT-2+	(used in Canada)	NW
Token Ring	Method for distributing information in networks	NW
802.11 IR	802.11 Infrared Protocol	NW
ISDN	Integrated Services Digital Network	NW
DECT	Digital European Cordless Telecommunications	NW
GSM	Global Systems for Mobile Communications	NW
FHMA	Frequency Hopping Multiple Access	NW
CDMA	Code Division Multiple Access	NW

Table 7.1 Popular Protocols Chart (continued)

Acronym	Full Name	Used For[1]
Wideband CDMA	Code Division Multiple Access	NW
Composite CDMA/TDMA	Combined for Code and Time Division Multiple Access	NW
TDMA	Time Division Multiple Access	NW
DCT based TDMA	Time Division Multiple Access	NW
DCS based TDMA	Time Division Multiple Access	NW
REFLEX	Two-way paging	NW
FLEX	One-way paging	NW
inFlexion	Supports communications in 50kHz narrowband channels for PCS	NW
Hybrid CDPD	Hybrid Cellular Digital Packet Data	NW
CDPD	Cellular Digital Packet Data	NW

[1] SW–software; HW–Hardware; and NW–Network.

Section 2:

Making it Work: How to Plan, Design & Deliver Wireless Computing Solutions

Planning a Wireless Computing System

Defining Business and "User" Requirements

Whether you are considering using wireless computing for individual use or for a group of mobile workers in the field, it is important to conduct thorough and proper planning. Individuals and small groups (10 or fewer mobile workers) may elect to take a horizontal approach by selecting existing products which have been proven to work together. If you are planning a wireless computing system that will become a part of an enterprise system or will significantly impact a company's operations, it may be advisable to take a vertical approach and develop a customized solution.

At ACT, Inc., we have generated a five step approach that we follow when developing a wireless computing solution. It is called the Project Planning and Life Cycle. This procedure can be applied to the development of any information management system, particularly those which employ new and emerging technologies. The steps for our Project Planning and Life Cycle approach include:

- Business Case Development
- System Design
- Pilot Development and Analysis
- System Development and Implementation
- System Installation and Rollout

Business Case Development

The first step in developing a business case is *Project Definition*. During this phase one must identify the business requirements which the wireless computing system is expected to satisfy. For example, is the company seeking to increase inventory turn or improve the quality of customer service? Once the business requirements have been clearly identified, measurable goals and benefits to be achieved from the wireless computing system should be set.

Goals may include changes in current operations that will enhance quality. Since many companies have implemented Total Quality Management (TQM) programs, these goals are often easily identified. The challenge is achieving them. Goals may also include financial returns. Many wireless systems (e.g., dispatch and point-of-sale) allow the consumer to perform credit verification on the spot or process invoices faster. The result is a reduction in accounts receivable and days receivable. Gaining a marketing advantage over the competition will also drive goals. If a salesperson can provide critical information while meeting with a customer, they may have a better chance of closing the deal on the spot. In some cases the marketing advantage has become a competitive requirement. Federal Express and UPS have raised the bar for customers' expectation of service in the package delivery industry. Other companies are now moving to implement systems which will allow them tell their customers the exact status of their shipment.

The goals must be realistic and attainable given the existing technology. Remember, it is easy to become so enamored of new and emerging technologies that expectations may exceed what is actually attainable. The benefits which are defined must obviously satisfy the business requirements. This basic step often does not receive the attention that it deserves in the systems planning process.

Make your goals meaningful by giving them measurable targets. For a goal to be of value, you have to make it count. It is possible to tie the benefits of wireless computing directly to the income statement or balance sheet. Productivity improvements are often realized in the jobs or tasks that field employees perform and how much resource they consume in performing these jobs. I call these *Job Productivity* and *Performance Productivity*. Job productivity measures may include reduced time in responding to customer assignments, an increase in the amount of revenue earned from each technician, or a reduction in overtime. Average-revenue-per-job and weighted salary factors can be used to establish job productivity measures. Performance productivity measures may include a reduction in the amount of time spent performing a job, a reduction in the ratio of in-house staff to field employees, or less time preparing reports. These measures can also be directly tied to revenue attainment. In addition to the improvement in the income statement, wireless computing systems can positively impact the balance sheet as well. Asset and Cash Flow improvement measures are often achieved when reductions are realized in accounts or days receivable. See Appendix I for a list of measurement formulas.

Once the goals and benefits have been made clear, the systems planner should identify the technologies able to support them. The methods and steps for implementing the identified technologies should also be defined. In most cases, implementation of wireless computing technologies will have a direct affect on the company's operations. In addition to identifying the enterprise computing systems which will be affected it is important to think through the effects that implementation will have on daily operations. What will it take to install and test the wireless computing system? How will mobile users be trained? Moreover, how will this be achieved without severely disrupting daily operations?

Finally, as the project is being defined take care to determine who is required to make it a success. Since the development and implementation of a wireless computing system requires the coordination of several parties, it is essential to gain proper staffing at the outset of the project. The project team should include management, mobile workers, and someone to represent those affected by the system.

Obtaining management buy-in, from all levels, is important. Depending on the magnitude of the project, the initial outlay for a wireless computing system could be substantial. Proper coordination should involve several departments within the organization. Not only is it smart to gain the endorsement of a senior level manager who has the authority to make the necessary budget approvals, it is prudent to gain concurrence from managers at all levels in the implementation chain. That is, those managers who can facilitate the implementation and acceptance of the system.

Mobile workers play a pivotal role in the system planning process. It is they who become the mobile consumer. It is wise to involve mobile workers early in the planning process. They can assist in assessing the viability of developing a wireless or mobile computing solution. The mobile worker is also pivotal in determing the feasibility of implementation. Mobile workers should be canvassed to determine their needs and desires. One or more representatives should be selected to partcipate on the planning team. By involving the mobile worker in the planning process, you will have an ally to assist in gaining acceptance of the mobile device. Without acceptance from mobile workers the effort required to maximize the benefits from a wireless or mobile computing system is dramatically increased.

Last but not least, be sure to involve those who are affected by the system. These are your customers, business partners, and fellow employees. If your mobile device will be used to collect information from customers, make sure that the customer is willing to participate in the collection process. A major package delivery company determined that customers needed to see their signature when they gave it, and the physical platform had to feel as close to pencil and paper as possible. Combined with the level of durability required, all pen-based devices available at the time were eliminated. Insurance salespeople found that customers felt intimidated when computers were used to collect information to provide policy projections on the spot. They felt an invasion of privacy. Managers sought devices to perform comparable functions that did not "look" like a notebook computer. Remember to verify that the wireless system will be accepted by the customer, especially if they will be expected to interact with it.

Concurrence from your business partners and fellow employees may also be necessary. Will the wireless computing system be perceived as a threat by your resellers? Will distributors need to allow merchandisers special access to facilitate collection of inventory information? Do fellow employees' activities need to be altered to successfully implement your system? Are others affected whose consent will facilitate deployment? Determine the role that others will be expected to play or how they could affect deployment. A major shipping company developed a state of the art wireless computing system only to find that rollout was being sabotaged by their union workforce. Mobility and instantaneous access to information can affect others. Take steps to ensure that they become a partner in the system rather than an adversary.

Project Requirements

The final step in the Business Case Development is determining Project Requirements. Once the Project Definition has been completed and the staff selected, defining the project requirements will be easier. Several areas need to be evaluated including:

- Mobile Interface
- Response Time
- Information
- Security
- Form Factor
- Weight
- Communication Points

The *mobile interface* includes the software view and the hardware input mechanism that the mobile consumer interacts with to use the mobile device. The software view may include windows, screens, lines of text, or simply pictures that direct the mobile consumer on how to interact with the device. The hardware input mechanism may be a keyboard, pen stylus, bar code wand, buttons, or radio modem which functionally allows the mobile consumer to communicate with the software. The mobile consumer representative on the project team will be extremely helpful in the selection of the mobile interface. He or she can quickly assess the viability of the interface and the functionality of the mobile device. Remember, it is this person who is performing the activities on a daily basis. The mobile consumer should be integral in selecting between the number of successive screens; the layout or number of windows; input from a keyboard, pen, button, or bar wand. This project member can also help determine if backlighting will be necessary. Rather than select the device at this point, the acceptable options for the mobile interface should be defined and listed.

A wireless computing system is a tool which facilitates performing a given activity. The focus will, therefore, be on that activity, not on the performance of the device. To maintain focus on the activity, the *response time* must be reasonable. Reasonable can be considered 20 seconds or more for activities that are not rushed. On the other hand, a major stock exchange determined that the end-to-end response time had to be 6 seconds or less if traders and others working on the exchange floor were expected to use it.

Response time plays as important a role in gaining consumer acceptance as the mobile interface. Response time is not measured by the speed of the processor, the amount of memory or the speed of the wireless network. It is measured by the time from which the mobile consumer initiates a request by making contact with the mobile device until the response is received. To minimize response time, all components of the wireless computing system must be tightly integrated and fine tuned. Rather than determine the speed of each component, however, the project requirements process should merely determine the end-to-end response time that the wireless computing system must achieve.

The project requirements document should also include all *information* required to fulfill the goals of the wireless computing system. This information is not necessarily limited to the mobile data. It may include information that resides in host computers or other systems that are necessary to complete processing of the wireless computing system. For example, the mobile consumer may send a request to the host system to calculate an average price and return the result. In this instance the information would include data in a price database that resided on the host computer. Most of the information collected at this step will be used to complete the Mobile Data Definition Chart.

Access to time-critical information is usually limited to those who have a need to know. The proper level of security is necessary to restrict access. The level of *security* required may vary along the flow of information. Certain information is of no value until it is combined with additional information or placed in the hands of someone who knows how to use it. The project requirements document should determine the level of security which is required at each step in the flow of information.

The mobile consumers' ability to carry the mobile device and use it while performing a given activity is governed by the *form factor* and weight of the device. The shape of the device must allow it to be easily held or worn on the body. Belts, shoulder harnesses, straps, cases, and other gear are often used to make the device easier to carry. The shape of the device will also determine how effective or durable the antenna is for wireless communications. Many antennas must be able to move freely and change directions to increase reception by the wireless network. The antenna must also be easily stored or retracted to avoid damage while the device is being carried. The weight of the device will determine the mobile consumer's ability to carry it without fatigue. Many mobile workers carry other items such as packages, briefcases, or tool belts. If the mobile consumer is expected to carry the device throughout the day, it generally cannot weigh more than two pounds. The weight is not as important, however, if the mobile activity allows the device to be mounted in a vehicle. The project requirements process must determine the form factor and weight which is necessary to allow the mobile consumer to use the device.

One of the key advantage's of wireless computing is that the mobile consumer can receive, collect, process, and disseminate information when and where they need it. To enable this, a wireless computing system must encompass all *communication points*. The project requirements document must list the location and characteristics of all possible communication points. For the mobile consumer these include the geographical locations and whether or not they are in a building or a vehicle. For stationary points within the wireless computing system, the list should include the description and location of computers, docking stations, and other physical components from which communications must take place.

Project Requirements

Criteria for selecting a wireless computing system:

1. Can the company's business requirements be fulfilled with:
   - wireless network coverage and availability
   - applications software for the mobile device
   - usability of the portable device
   - integration with the enterprise system
2. Do productivity gains and the mobile data's time-value of information exceed the costs of acquisition, implementation, and maintenance (i.e., recurring network and support costs)?
3. Is it worth it?
   - What savings and increased revenue will result from productivity gains?
   - What is the time-value of information?

- How much will it cost to purchase or develop the hardware, software, and initially subscribe to the network?
- How much will it cost to install the system, train mobile users, and establish support?
- What are the recurring network fees?
- How much will it cost to maintain the software, hardware, and network connection?
- How much will it cost to provide on-going training and operating support for mobile consumers?

4. Is full implementation operationally feasible?

Designing a Wireless Computing System

The Mobile Consumer Is the Key

From software to hardware to communications there are numerous technologies which can comprise a wireless computing system. Wireless communication options include packet radio, CDPD, and specialized mobile radio. Portable hardware choices include notebook computers, personal digital assistants, and rugged handheld devices. Software is increasingly being ported to portable platforms and new mobile software is emerging. There are a myriad of technologies and products on the market which can become components of a wireless computing system. Each system component must work together seamlessly and efficiently for successful implementation. So, the initial challenge for the systems designer is what to select. Which products will work together? How can they be integrated to efficiently provide the functionality and performance that meet the specified requirements? To achieve maximum use of your system design, start with the person who will be using the system—go to the mobile consumer.

The mobile consumer is the person who ultimately makes the wireless computing system successful. He is the computer repair technician. She is the utility worker. They are the product merchandisers, the salespeople or the traveling executives. The mobile consumer is the end user of a wireless computing system.

He or she can make or break the success of the system. To achieve maximum results from your wireless computing system it is essential to involve the mobile consumer throughout the development process—from planning to rollout.

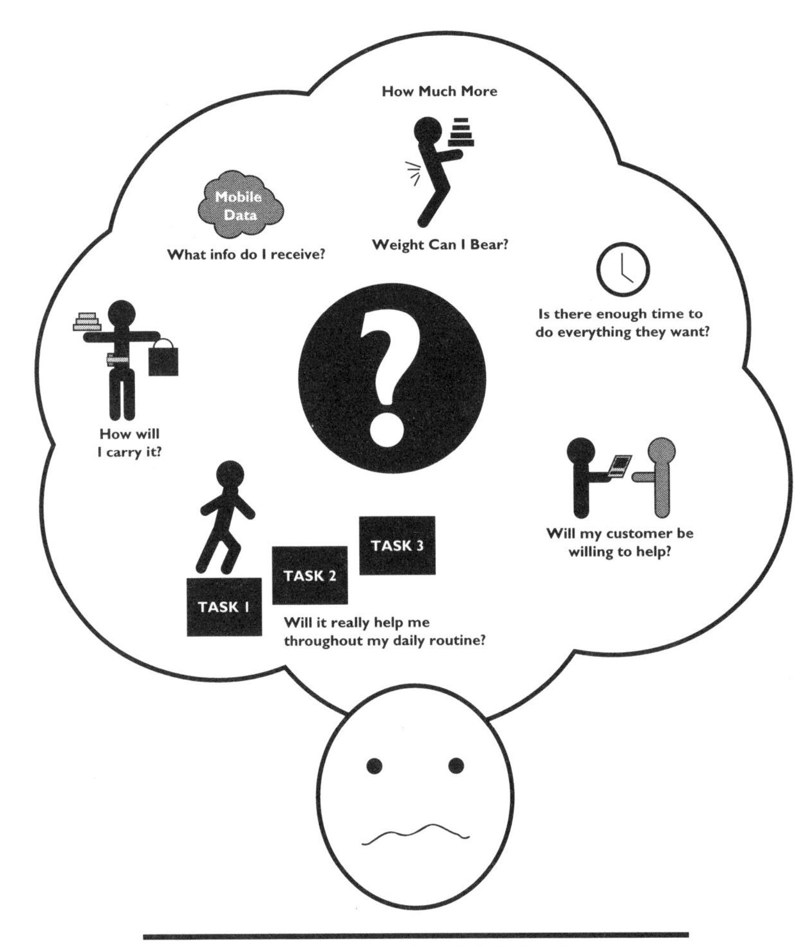

Figure 9.1 Design for the Mobile Consumer.

As companies begin to understand the real business benefits of wireless computing, many are taking the steps to learn how to develop and implement systems. Portable computing and wireless communications bring challenges not faced by users of desktop computers and landline communications. Power management is crucial in wireless computing, particularly as resource-intensive software functions and high-function peripheral devices are added. Transmission management requires attention to maximize the receipt and integration of mobile data while minimizing network costs. Providing mobility to employees and imposing new procedures may require input and approval from labor and legal departments. The mobile consumer can assist in streamlining these issues.

The mobile consumer can quickly assess the viability of the interface and the functionality of the mobile device. Remember, it is this person who is performing the activities on a daily basis. It is impossible to duplicate mobile requirements while sitting at a desktop computer or even by moving about within a confined, controlled environment. The mobile consumer can assist in selecting a viable interface. This person should be integral in selecting between the number of successive screens; the layout or number of windows; input from a keyboard, pen, button or bar wand. He or she should also be involved in the final selection of the portable device. This person can assess the "portability" of the device: where the antenna is positioned, what options (i.e. straps or holsters) will make it easier to carry; or the feasibility of carrying extra battery packs. By involving the mobile consumer, you will have an ally to assist in gaining acceptance of the mobile device.

As you prepare for the pilot and rollout, the mobile consumer can help you determine the training, maintenance and support requirements to ensure success. By involving someone who is selected by their peers, you may increase acceptance of the system. Many have encountered roadblocks when trying to rollout systems to field service groups who felt threatened. On the other hand, other companies have surpassed their expectations by showing the mobile consumer how the system would benefit them.

As more companies move to implement wireless computing systems, the benefits to be realized will surely increase. Careful planning is required to stay abreast of emerging technologies and reach targeted business goals. The system design is critical in achieving real success and a smooth roll out. To create the best design, remember to focus on the mobile consumer.

The Selection of Components: Build or Buy?

A major decision which must be made in the initial stages of designing a wireless or mobile computing system is whether to build a vertical application with custom components or buy off-the-shelf components to construct a horizontal application. Building a vertical application will allow you to customize the system to the unique requirements of the mobile consumer. This requires more time and money. Software development is necessary to provide the applications and systems functionality for the application. Hardware modifications may be necessary to make the mobile device more portable. Sufficient training and support must be put in place to facilitate smooth operations using the mobile system. A vertical solution, however, will provide the end-to-end functionality required to maximize the benefits to be gained from a wireless or mobile computing system.

As more products hit the market, horizontal applications become a viable alternative. In addition to ensuring that components selected for a horizontal application will provide the required functionality, it is imperative that all components work together.

Standards have not yet evolved to ensure 100% compliance between components. Consumers have not used mobile computing solutions long enough to readily determine if user interfaces and applications will provide the ease of use and functionality that justify computing "on the go". Data transfer from mobile devices to desktop computers is possible using off-the-shelf software, cables, and connectors. Docking stations that facilitate this process, however, have only recently begun to hit the market and are still evolving. Wireless networks require different software and modems than those which are used to communicate via landline networks. Some components, such as software and PC cards, consume much more power than others. Despite these issues, wireless and mobile computing systems can be constructed using readily available products.

When purchasing components for a horizontal wireless or mobile computing system it is important to select them carefully. Take time to make sure that all components work together to perform the functions that you require. With the exception of MS-DOS- and Windows-based laptop and notebook computers, most products have only been on the market for a few years at most. Many components are not yet plug-and-play. Verify that the software versions will perform on the hardware models that you select. Validate the compatibility of PC cards and other peripheral devices. If you are using a wireless network, make sure that you have the software and hardware necessary to communicate. The distributor or retail store which sells the components that you desire can check with the manufacturer or software vendor to determine the level of compatibility.

Consider what may be required to update the components in three to five years. Does the software vendor have a reputation for building new products which are backward-compatible? Will you dispose of the hardware, or are sources available to upgrade it with the latest technology? Although products enhancements are a continuing phenomenon among computer products, the useful life of most wireless and mobile computing systems is three to five years. Spend time evaluating and choosing the software, hardware, and networks that you purchase. Careful selection of components used in a wireless or mobile computing system will help to avoid problems in use down the road.

Since not all mobile software and hardware are as intuitive as you may expect, review all product screens and literature, attend demonstrations or, if possible, test the products yourself before purchasing them. In addition to determining whether or not the software performs the functions that you desire, make sure that the interface allows you to physically perform the functions in the mobile environment. For example, can the functions be performed with a single keystroke or a mouse or, must multiple keys be pushed at one time? Many portable devices are difficult to operate when using them in cramped environments like an airplane seat or when holding them in one hand and operating them with another. Pushing multiple keys at once may prove awkward. Are the screens easy to read in poorly lit areas or bright, sunny locations? The amount of information displayed on a screen or its shading can determine how easy it is to read as much as backlighting. This is particularly true for portable devices with small screens. Can you perform the functions that you desire with relative ease or must you navigate through several screens or steps to complete each process? When you are on the go it is not always feasible to spend a great deal of time on routine functions. To achieve true mobility you must be able to perform most functions, particularly routine functions, quickly and with relative ease. Whenever possible, test the mobile software and hardware to ensure that it is easy to learn and use.

Messaging software and services have emerged that offer consumers wireless computing solutions which are partially or completely turnkey. That is, when consumers purchase products or subscribe to services that provide messaging, everything will work together without modification or additional development. Wireless messaging software provides E-mail functionality over wireless networks. Basic messaging functions such as sending messages, attaching files to messages, and receiving confirmations are available. These software packages can also interface with local area networks to allow the mobile consumer to use enterprise messaging systems while traveling. They also employ transmission cost management functions to assist the mobile consumer in controlling the cost of sending messages over wireless networks. Wireless messaging services provide the software and interface hardware necessary to perform store and forward messaging over wireless networks. The service providers offer Internet and paging access as well as facsimile transmission for the mobile consumer. Message center operators are available to provide support. Messaging software and services provide a relatively simple and beneficial form of data communications for mobile consumers.

Eventually most people using mobile computing devices will return to their home or office. Data from the mobile device may need to be transferred to a desktop or host computer. Off-the-shelf software and hardware interfaces are readily available for data transfer. You may also require extra batteries or a recharging unit to use at your home or office. Remember to back up your data. Portable hard disk and tape units are available, or you may choose to back up your data upon returning to your home or office. Determine your computing requirements as you travel from home to office. This will help you to purchase the components that you will require for your horizontal system.

Once you have determined the components required to complete your horizontal wireless or mobile computing system, reassess your goal to verify that your purchase is cost-justified. We so often become consumed in the marvels of technology that it is easy to lose sight of the original system objectives. Moreover, additional benefits are sometimes realized during the process of selecting components. Determine the use-value of the wireless or mobile computing system that you have selected. That is, will you realize the expected benefits as you actively use the system? Keep in mind the functions that you will be able to perform and how you will benefit by executing them as you move about. Most people who use wireless and mobile computing systems can't imagine how they ever got along without them.

Mobile consumers who are part of a sizable group (i.e., 10 or more) and who perform activities which are essential to the operations of group may find that horizontal solutions are not adequate to achieve the functions that they require. Software or hardware adjustments may be needed to allow the portable device to conform to their unique operating environment. They may require customized screens or functions to closely resemble or duplicate current operations. Applications may need fine tuning to ensure operations over a full shift of 10 hours. Hardware may require adjustments or special packaging to make it easier to carry while performing other activities. When off-the-shelf products do not meet the needs of the mobile consumer group a vertical application may be justified.

If this is the case, more focus should be placed on the system planning process starting with the system design (Figure 9.2).

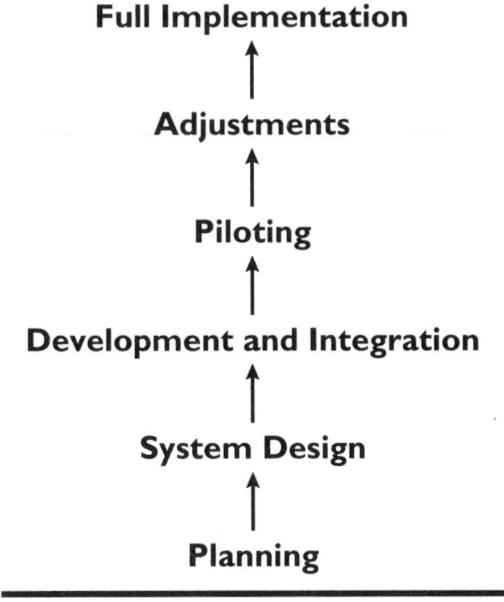

Figure 9.2 System planning process.

System Design

Of paramount importance in system design is protecting the integrity of mobile data. To gain confidence from the mobile consumer, mobile data must be complete, accurate, available and secure. All data that is transmitted must be received exactly as it was sent. Mobile consumers must be able to send data *when* it is needed. The system design should provide the assurance that mobile data is accessible only by those who have a need to know. The system design should ensure the reliability of mobile data while meeting the needs of the mobile consumer.

After consulting the mobile consumer, the systems design process should begin with the selection of components. There are four major components which comprise a wireless computing system:

- Host Computer

- Software
- Wireless Network
- Portable Device

Host Computer

Wireless computing allows the mobile worker to receive, collect, process, and disseminate information when and where they need it. For this information to be of value to the corporation, it must be received from or communicated to enterprise systems. The host computer is the information repository within the enterprise system. It could be a mainframe, minicomputer or even a desktop or portable computer. The host computer may be maintained by a MIS department, service bureau, the network carrier or the mobile consumer. Timely and accurate communication between the host computer and the mobile consumer gives the mobile data value.

The host computer plays an essential role. It houses the applications which assist the corporation in managing its business. Inventory management, order processing, asset management, and product pricing are a few of the applications which may reside on the host computer. When integrating a host computer into a wireless computing system, give careful consideration to:

- Applications
- Communications interface
- Operations Procedures

The host *applications* which the mobile consumer will communicate with must be identified. The designer should define the information items, or data fields, and their frequency of update. Data field definition includes everything that's necessary to read and write to the host files including the field size and format. The frequency of update should include how often the information is available to be updated and how often updates from the mobile consumer can be processed. Using the Project Definition, the systems designer can schedule and synchronize the updates. This is a critical step in establishing the value of the wireless computing system and also in managing its costs.

Information updates will not take place without the proper *communications interface*. These interfaces consist of hardware that provides the conduit, and software that provides the "handshake," for information to be sent between devices. The systems design should define the communications interface(s) available at the host computer. Available interfaces may be directly connected to the host computer or may be connected through a front-end processor. They may require 3270, AS400, or simply a serial protocol. Since host computers are usually designed to talk to landline communications systems, they are often session based. That is, a dedicated path or bridge is established between the two points of communication. This path or bridge remains intact until the communication has been completed or the session is over. Paths or bridges may be divided between several communication points. Any two points of communication will have a dedicated pathway. The path or bridge must be managed and paid for as long as the communication takes place. Most wireless technologies are based on sessionless communications. Information is disassembled, grouped into packets, and sent to its destination. The communications path is shared among several communication points. The information groups or packets being sent are managed and paid for. The pricing of communications is usually not based only on time and volume; however, these are two factors which will likely impact the speed and cost of transmission. When designing the connection to the host computer, therefore, it is not technically and financially advisable to extend the host communications method to the mobile device. In most applications it is advisable to perform a communications translation *before* the data reaches the host computer. The systems designer should consider this when defining the available communications interfaces.

Many host computers are part of an enterprise system which has been in operation for a considerable period of time. As we noted earlier, the applications on these systems assist the corporation in managing its business. Procedures, formal or not, exist to facilitate smooth operations. The system designer should take care not to rock the boat. Existing *operations procedures*, therefore, must be taken into consideration in the system design. Communication to and integration with host applications need to adapt to existing technical and security requirements. This must happen without compromising the objectives and timeframes of the wireless computing project.

Software

Many are attracted to wireless computing after seeing the latest "state-of-the-art" portable device or hearing about the wonders of modern day wireless communications. The natural inclination is, therefore, to begin investigating wireless networks or portable hardware. Thorough integration of wireless networks and portable hardware is essential for a successful wireless computing system. While the companies who provide these products and services have made, by far, the largest investment in wireless computing, the selection of the network or hardware **should not** be the first step. Rather, the initial focus should be on selecting the software which will drive the application.

Software resides on the host computer as well as the portable device. The software on the host computer is either an extension of or addition to the enterprise system. Our focus shall be on the software which resides on the portable device. Three categories of software must be considered during the selection process:

- Applications Software
- Communications Software
- Operating Systems Software

As the essence of computing systems, the *applications software* must ensure that the system can deliver the functionality to meet planned requirements. Whether it is purchased or developed for a specific application, the following criteria should govern the selection of applications software:

- Functionality
- Ease of Development
- Performance
- Scalability
- Upgrade Path
- Support

Above all, the applications software must perform the functions necessary to empower the mobile consumer. Whether it is searching for information or performing calculations, the applications software must do what is needed to satisfy the specifications of the Project Requirements document (see Chapter 8).

If the decision is made to purchase completely developed, or shrink-wrapped, software, the ease of development is not a relevant consideration. There are some E-mail and dispatch packages which run on portable hardware platforms and perform well with selected wireless networks. The selection of shrink-wrapped mobile software, however, is limited. Since most applications software for mobile devices requires partial or complete development, ease of development must be considered. The complexity of the application, software programmability, and the pool of available programmers are major factors in determining the ease of development. The software should provide programming links or calls to other software including applications programming interfaces (API). APIs for wireless communications let the programmer improve communications performance on wireless networks while managing the transfer and cost of mobile data.

Mobile applications software must run on portable devices with limited computing power and resources. This software must provide acceptable performance with slower processors, modest memory and limited battery life. While all mobile software takes these issues into account, some packages perform better than others. Both pre-packaged and customized application software must demonstrate performance acceptable to the mobile consumer.

Many portable devices have hit the market and even more are to come. Personal digital assistants, palmtop, and notebook computers are a few of the portable devices available to the mobile consumer. They come in many sizes and support several platforms. Several companies who have implemented wireless computing systems have found that it is virtually impossible to select one platform and device for all mobile consumers. Moreover, not all mobile consumers will need the same selection of applications. It is often necessary, therefore, to run mobile applications software on more than one platform. Applications software must be scalable. That is, it must be able to be easily moved from one portable platform to another.

As a result of business demands and technological developments, the life span of wireless computing systems may well be shorter than the initial planning and development cycle. After the successful rollout of a wireless computing system, the benefits are often realized by many. Mobile consumers and managers often want to increase functionality as well as the amount and frequency of mobile data. When new products hit the market, support for older models or versions is eventually withdrawn. To guard against prohibitive enhancement costs in the future, it is imperative to ensure that an upgrade path will be available. An upgrade path means more than being upwardly compatible with planned future versions. It also means being offered future versions with the additional functionality that is desired, being able to move to new mobile platforms, and having adequate training and support available during the upgrade process. In short, an upgrade path for applications software means having everything that's needed to install and use future versions in the next-generation wireless computing system.

The applications software you select should have all the tools and services necessary to make it easy to install, learn, and maintain. These tools and services include written and on-line documentation and tutorials, technical hotline, train the trainer classes, classes for the mobile consumer, custom development services, and on-site assistance from local offices or value-added resellers. The amount of support which is required is dependent upon the ease of use, functionality, and adaptability of the applications software. Adequate support is essential for the successful development and rollout of a wireless computing system.

Communications software determines the quality and efficiency of sending mobile data over wireless networks. It can be provided by the network carrier, applications software vendor, or by the independent software vendor who developed it. Because landline communications employ different technologies than wireless communications, software that manages the wireless communications functions must be designed and written to exploit the unique advantages of wireless technologies. Given the number of wireless technologies and communication protocols, the software should be modular and adaptable to current and future standards and systems.

Communications software can be shrink-wrapped and sold for specific applications, such as E-mail. It could be bundled with applications packages such as contact managers or field dispatch. Or it could be sold to software developers to allow them to wirelessly enable the applications they create. Whether the wireless communications functionality is delivered from an applications software package or an applications programming interface, the following criteria should be considered when making a selection:

- Performance
- Wireless Compliance
- Interoperability
- Efficient Protocols
- Upgrade Path
- Support

As is the case with applications software, communications software must operate within the confines of the portable computing platform and at a speed that is acceptable to the mobile consumer. The performance of communications software is largely governed by the speed at which it processes data for transfer over the wireless network and the resources (e.g., battery power) that it consumes during performance.

Wireless compliance is determined by how well the communications software functions with the technologies of the wireless network. The data compression rate should be high to minimize the size of the data items (e.g., packets) that are sent. In packet networks, the software should optimize the grouping and sequencing of packets to increase the probability of successful and one-time transmission. Checkpoint restart, recovery, error detection, and correction methods will help ensure complete and accurate data transmission. Unlike landline communications, wireless networks are subject to latencies, which are created from external forces (e.g., wind, rain). The communications software should provide latency management to guard against data loss when external forces impair the wireless path.

Several protocols and environments may be used to provide end-to-end communications in a wireless computing system. To facilitate smooth communications, the software should employ protocols that can be easily transferred or translated for use in different environments. This provides the interoperability that eases development and improves performance between different environments. This is not only important to enable communications with wireless networks, it is also necessary to communicate with different host computers.

The protocol which is used by the communications software helps determine how quickly and how much information is sent over the wireless network. The overhead of the protocol can increase the volume of data that is sent. The operation of the protocol can increase the number of transmissions which must take place to transfer data. Efficient protocols have minimal overhead and require fewer transmissions to send data.

As with applications software, an upgrade path for communications software means having the tools and services that are needed to install future versions in the next generation wireless computing system. The next generation may include not only different hardware environments, it may also include different wireless networks.

Shrink-wrapped communications software shares the same upgrade issues as applications software (see Section II, Chapter 9—*Applications Software*). The upgrade path should offer everything that's needed to install and use future versions in the next generation wireless computing system. If the communications software is bundled with the applications package, these issues should be addressed by the vendor who develops and sells that package. For developers who elect to use application programming interfaces (API), this software should be modular and portable. Modularity will allow new protocols and methods to be added without a major software rewrite. A portable API does not consume much storage and memory and is written in a language (e.g., C) that has been ported to several environments. Portability will allow the developer to move applications between different platforms more easily. There are several wireless networks to choose from and even more portable devices. Competition and varying application requirements virtually ensure that there will always be choices. To avoid being tied to a particular platform or network, make sure that the communications software has a flexible upgrade path.

Shrink-wrapped communications software shares the same support issues as applications software. For applications programming interfaces (API), the critical support issues are how well the software is written, how well it is documented, and how quickly and accurately technical questions are answered.

As with applications software, the challenge of *operating systems software* in portable environments is to optimize the functionality which can be placed in a platform with limited resources against the ease-of-use of the system interface. Although some portable devices support, proprietary operating systems or hybrids of popular ones, this is quickly changing. Many keyboard and button-based portable devices support or are moving to support, MS-DOS, Windows, Windows 95, Magic Cap or GEOWORKS. Some handheld and other portable devices support hybrid versions of DOS. Pen-based portable devices usually support PenDOS, Windows for Pens, PenPoint, PenGEOS.

As is the case with all software, a major concern with operating systems software is that it will continue to be supported and applications will continue to grow. Unlike desktop computers, a single standard operating system (O/S) does not exist for all portable computers. Moreover, it will take several years, at least, for a single standard O/S to establish itself. Unless provisions for ongoing support have been made and the operating system offers unique and overwhelming advantages for the application, a "non-standard" operating system should not be selected. In this case, a standard operating system is one that is generally available and has been selected to run on hardware platforms of several vendors.

If the decision has been made to purchase applications software, the choice of operating system is usually relatively straightforward. The selection is limited to the operating system(s) which the application software runs on. If there is more than one operating system to choose from, the decision should be guided by the mobile platforms that fulfill the project requirements. That is, consider the operating system which runs on the portable hardware platform that the mobile consumer is most likely to use. If the decision has been made to develop custom software, the operating system should facilitate development. The Project Requirements document will serve as a guide to determine the acceptable hardware platform(s). The operating system will help the applications developer to maximize the value of the software that must be coded while optimizing the performance of the hardware. Especially when developing a custom system, the criteria which should be considered when selecting an operating system include:

- Functionality
- Programmability
- Performance
- Multi-vendor Interoperability
- Support

First and foremost, the operating system must provide the functions necessary to manage the resources of the wireless or mobile computing system. As we discussed in Chapter 4, there are several mobility functions that can enhance the performance of mobile systems. The interface should be adaptable to the application requirements and easily learned by the mobile consumer. Peripheral devices such as PC cards, radio modems, and bar code wands should be attached or inserted and removed without disrupting operation of the application. The operating system must support the basic functions required to support the application computing requirements of the mobile consumer.

It is virtually impossible for an operating system to provide all of the functions that are needed for every conceivable application. This would require a size and complexity beyond the capacity of most portable computers. To allow other software to extend the capabilities of an operating system, it must be programmable. Independent software vendors and custom developers need the flexibility to invoke or call operating system functions to increase the functionality and value of application programs. This may include the dynamic re-allocation of system resources such as adding memory or changing device drivers for PC cards.

The custom software must be written with elegance and efficiency to maximize its performance for the mobile consumer. The operating system must provide a high level of performance to accomplish this. It must be ported to run on the fastest processors, handle all available memory, and perform all functions at maximum speed. The operating system must support a level of performance that satisfies the patience and matches the abilities of the mobile consumer.

The selected operating system must provide a level of assurance that the wireless or mobile computing system will continue to operate as expected. Multi-vendor operability is one means of assurance. Hardware vendors invest substantial engineering, financial, and marketing resources when they decide to support an operating system. Their selection of an operating system is usually a long-term commitment. Knowing that several vendors offer the selected operating system on their hardware platforms provides this assurance. A high quality and adequate level of technical support provides additional assurance. Technical support is measured by the expertise and availability of the support staff as well as the quality of the documentation.

The selection of an operating system is fairly academic when the decision has been made to purchase ready-made, or off-the-shelf, applications software. Most wireless and mobile computing systems for individuals or small groups will be based upon applications software that already exists. This is particularly true when notebook computers, palmtop computers, or personal digital assistants are used as the hardware platform. If the decision is made to develop a custom application, however, the selection of an operating system deserves more attention. In this case, the selection goes hand in hand with the evaluation of the portable hardware. Use the aforementioned criteria to select an operating system that provides the foundation for a functional and reliable wireless or mobile computing system.

Wireless Network

The selection of wireless networks that handle mobile data efficiently is somewhat limited. But that is quickly changing. As we have learned, several carriers are deploying or readying their networks to efficiently handle the two-way transmission of mobile data. As the competition intensifies, carriers are taking steps to differentiate and increase the value of their services. Alliances with Independent Software Vendors, Systems Integrators, and Resellers have created software packages, services, and end-to-end systems that incorporate transmission over wireless networks. The result is an emerging selection of systems for the mobile consumer.

Carriers are responding to the first step in the selection of a network for a wireless computing system—assessing the availability of software and hardware.

Many mobile consumers, particularly those using systems for personal productivity, use wireless computing systems for electronic mail, or E-mail. Others only want to send the output from software applications. Since it is less expensive to purchase existing products than to develop new ones, the selection of a wireless network should start with the identification of compatible software and hardware products that fulfill the application requirements.

The selection of a wireless network is obviously not limited to the availability of network-compliant software and hardware. A growing number of software and hardware products support multiple wireless networks. Many projects require development of a custom wireless computing solution. As with other components of a wireless computing system, the selection of a network should begin with the Project Requirements document. (see Chapter 8—*Defining Business and User Requirements*). The network evaluation must be conducted based upon the needs of the business application. How much mobile data will be transmitted? How often will it be sent? Who will receive the mobile data? Where will it be sent from? Where will it be sent to? Whether you are evaluating existing networks or those planned for the future, the selection criteria remains the same:

- Coverage
- Availability
- Reliability/Speed
- Capacity
- Security
- Cost

The wireless network must provide the geographical coverage necessary to allow mobile consumers to access it. No wireless network offers ubiquitous, two-way communications service for mobile data. Available coverage is sufficient, however, to allow access by mobile consumers in most situations. Newer wireless networks (e.g., packet radio, CDPD) provide coverage in major cities and along highly traveled transportation corridors, reaching the vast majority of the U.S. population. Older networks (e.g., circuit switched cellular) provide coverage that extends into many remote areas, not densely populated. Since radio waves cannot penetrate all elements (e.g., specially coated windows), wireless networks are not always accessible from within buildings. Some wireless networks handle transmission from mobile consumers who are in motion better than others. The systems designer must determine the locations from which the mobile consumer will need to access the wireless network. This will assist prospective user in rating the coverage that the wireless network will provide for their system.

In addition to coverage, the wireless network must be available to the mobile consumer *when* they need it. Insufficient network capacity, faulty equipment, and other factors can reduce availability. Mobile consumers, or subscribers, may sometimes require multiple attempts before successfully transmitting mobile data. This consumes extra time, battery life, and patience—all precious commodities for the mobile consumer who places a high time-value on the information they need. Network availability is virtually impossible to assess without complex traffic studies or actual use. Network carriers generally commit time and resources to effectively manage their networks. This includes traffic analyses, which let them circumvent availability problems. A review of the carrier's growth and management plans will help to assess and rank its availability.

A great deal of attention is often focused on the speed, or *throughput*, of the network. The real issue, however, is how quickly mobile data can be successfully sent from one mobile consumer to another. That is, from the time that a mobile consumer invokes the request to send, until the mobile data arrives at its intended location. The network throughput is only one part of measuring the speed of sending mobile data. End-to-end speed is also affected by other components such as communications software, portable hardware port speed, the radio modem, and the number of times that the mobile data must be sent to arrive successfully. Error detection and correction routines in the network, modem, and communications software help to increase the reliability of the transmission. Remember that the throughput of the wireless network is not an adequate measure for the speed of a wireless computing system. The reliability and speed of the end-to-end transmission of mobile data must be assessed and ranked when evaluating wireless networks.

The capacity or volume of mobile data which can be sent in each transmission, will vary substantially from network to network. Network capacity will affect the amount of time required to send a given amount of mobile data from one location to another. Moreover, capacity will determine whether or not certain data files can be sent. Certain applications software packages (e.g., spreadsheet, graphics) do not allow files to be divided unless the file is broken up or separated by the applications software itself. It is often not operationally feasible to send files that must be processed by the mobile consumer before they can be viewed or used. Wireless computing systems should rely upon data compression algorithms to condense files into mobile data. The size of the files and subsequent volume of information which must be sent with each transmission may well limit the networks that can be considered due to capacity constraints (see Chapter 2). The systems planner must determine the maximum possible file size to measure the volume of mobile data that must be sent in a single transmission. This will serve as a benchmark for selecting the networks that offer the capacity required for the application.

Proprietary information is sometimes sent via wireless networks which must be restricted only to those with a need to know. Many networks employ encryption and other security measures to protect mobile data from access by unauthorized persons. Certain technologies and access methods contribute to network security. Spread spectrum technology, for example, is inherently secure. It is virtually impossible to intercept and reassemble data which is spread across several frequencies. Code division multiple access, CDMA, also requires that data be reassembled before it can be interpreted. All information is sent over the same frequency, however, on most networks. Providing security for mobile data should not be limited to the network. Rather, security should start at the time that the mobile data is created. When information is proprietary, it should be encrypted when it is collected or generated. Software on the mobile computing device and host computer should work in concert to protect mobile data. The design of the system should be such that the information can only be accessed and interpreted by those who have a need to know. This can be achieved in a variety of ways using software (e.g., automatic encryption, password protection). Remember, nature of wireless communications provides an opportunity for intrusion since mobile data is sent through the air. When mobile data is proprietary, security measures should be implemented throughout the system, not simply in the network.

Along with maintenance and support, the network is the major source of ongoing expense in a wireless computing system. The amount of recurring expense, or cost of usage, is often difficult to predict. Pricing, therefore, is a major consideration when comparing wireless networks. Here again the application requirements play a major role. Network pricing can be distance-, or usage-sensitive. That is, the prices can be based upon the amount of mobile data that is sent or how far it is sent. It should be noted that usage pricing does not apply to some wireless networks (e.g., LANs). This is the case for networks using frequencies that do not require an operational or frequency license from the Federal Communications Commission (e.g., spread spectrum). To use spread spectrum frequencies the *device* which accesses these frequencies must be licensed by the FCC.

The Wireless Network Comparison Chart in Figure 9.3 provides a framework for evaluating different technologies. It has not been filled in because the answers will vary with each application. End-to-end speed will vary with the configuration of software and hardware in the mobile unit as well as with the type and volume of information transmitted. Costs may vary based upon the usage pattern of the mobile consumer. At this point in the evolution of wireless computing systems, it is not possible to complete Figure 9.3 so that it will universally apply to every application. Likewise, the differences between the wireless technologies for each comparison item may not be well defined for all applications. A ranking scale of *high*, *medium*, and *low* is used to provide relative comparisons. Once it has been completed with rankings based on the requirements of a given application, the Wireless Network Comparison Chart may serve as a guideline for selecting wireless technologies. The purpose of this chart is only to offer direction. Additional technical and business issues must be considered when selecting a wireless network carrier.

The system design is a critical step in the successful development of a wireless or mobile computing solution. It must be driven by the needs of the mobile consumer. For individual consumers the system design is the careful selection of products and services that satisfy operational and mobility needs. For a large group or company, the design of a wireless or mobile computing system is usually a re-engineering effort which requires the coordination of several parties, skillful management, and fine-tuned integration. Thorough and efficient project management is required to pull all of the components together. When performed properly the system design will play a major role in the success of the system. It helps to control costs. It promotes the integrity of the system and of the mobile data. Finally, the system design facilitates effective deployment of the wireless or mobile computing system. While several other steps are of equal importance, time and resources must be invested to ensure a steadfast system design.

Wireless Network Comparison Chart

	Packet Radio	Cellular	PCS	SMR	Satelite	Paging
Coverage						
Availability						
Reliability						
Speed						
Capacity						
Security						
Cost						

Rankings H-High M-medium L-Low

Figure 9.3 Wireless network comparison chart.

Integration Issues

To maximize performance of the physical system, all components must be tightly integrated. This is very important. Wireless and mobile computing systems are composed of several components from different manufacturers. A set of well established, popular protocols that apply to all facets of wireless mobile computing systems does not exist. The complete and thorough integration of system components is essential, therefore, to ensure optimal performance.

Several areas require special attention in the integration process. Applications software should be coupled with communications software to ensure the smooth flow of mobile data. These elements should work together to make sure that the volume of mobile data does not exceed the network capacity. They can ensure that mobile data is sent in the proper format and at the preferred time of day. For example, certain data items which are not time-critical can be flagged or set apart for batch transmission when network traffic is low, or using networks which are less expensive. Many mobile consumers can easily access landline networks at the end of the day. Particularly with custom systems, the integration of applications software with communications software can serve to relieve the mobile consumer from having to make decisions on operating the system.

Applications software can also work with other systems software to help the mobile consumer manage peripheral devices. Since the mobile consumer focuses on screens generated by applications software, the screens provide an ideal opportunity to notify the consumer if a bar code wand is not properly attached, when a PC card should be inserted, or when the battery level is low. In most cases the operating system must provide the underlying functionality to allow this. Most hardware platforms, for example, require that the device be re-booted every time a different PC card is inserted. Operating systems which support mobility features, however, will allow PC cards to be "hot swapped." This means that new cards will not be recognized after insertion without re-booting the computing device.

Some systems software can enhance the functionality of the operating system. Communications software can take control of a serial port, for example, and manage the transmission of all mobile data sent through that port. This would allow the communications software to format and send the data so that it is recognized by the network that is supported by the modem connected to the serial port (Figure 9.4). The mobile consumer could be given screen notification of which modem to attach.

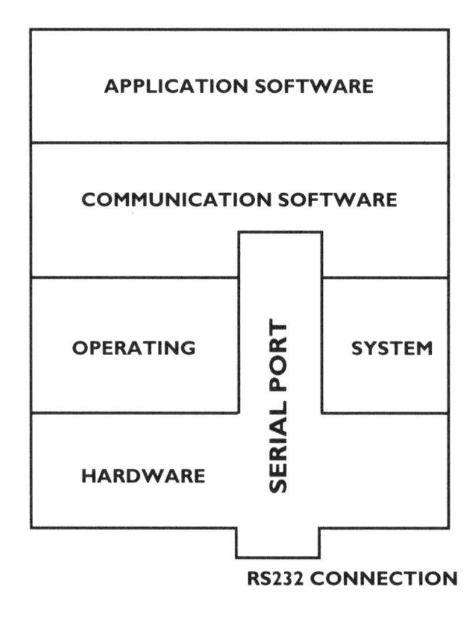

Figure 9.4 Application software, communication software and portable device.

Integration must also take place at the hardware level. Not all components are plug-and-play. Even if the interface and protocol are compatible, the components should still be tested. Hardware ports and peripheral connections are not always 100% compliant. For example, all Type II PC cards may not work in all Type II slots of hardware devices. The computing device and all peripheral components should be operated together to determine the battery consumption of the combined unit. Although PC cards, bar code wands, and other peripherals often have their own batteries, they may still consume battery power from the portable computing device. Whenever possible, all components should be assembled in a manner that minimizes battery consumption. Ideally, a single power source or battery pack should be provided to simplify changing or replacing the batteries for the mobile consumer. This means determining the amount of power that will be needed for a full day's operation and developing a battery pack that supplies that power. The battery pack must fit into the combined device, however, and the device must be altered so that all components draw upon the new battery pack. This is often only achieved in a customized hardware platform which has been re-engineered.

Making it Portable

Creating a customized hardware solution can become quite involved and expensive. Even if re-engineering of the motherboard and other internal components can be avoided, designing and producing a new casing and mold could cost $5,000 or more. A special production run using the new mold is even more costly. Unless minor modifications will be performed for 5,000 units or more it is probably cost-prohibitive to develop a customized hardware device. Furthermore, maintenance and support for customized devices is virtually nonexistent. It is always advisable to use products as they have been manufactured. For many applications, however, it becomes necessary to combine the computing device with peripheral devices to create a combined, fully functional mobile hardware solution. It is much easier to carry and operate a single unit rather than two or more components attached with wires or connectors. A single unit can be created from existing products by developing a semi-permanent attachment, or with carrying cases.

A semi-permanent attachment can be created using Velcro, glue, or screws. Some relatively lightweight peripherals (e.g., bar code wands) have been attached with Velcro. High-performance bonding glue has also been used. This is usually not a very elegant and reliable solution, however. The newly combined device is often vulnerable to the wear and tear of being carried around. It may also fall prey to weather elements such as rain, heat, and humidity. It is easier and often more reliable to attach peripherals to a computing device which has screw bosses. Screw bosses are threaded holes on the portable device where extra components can be semi-permanently attached by inserting screws. This allows peripherals to be attached at a lower cost and with more durability (see Figure 9.5). When devices are attached together attention must be given to the amount and distribution of weight. As a rule of thumb, a portable computer with all attachments should weigh seven (7) pounds or less to be considered portable. Handheld devices should not weigh more than three (3) pounds and the weight should be evenly distributed to make it easy to hold and operate.

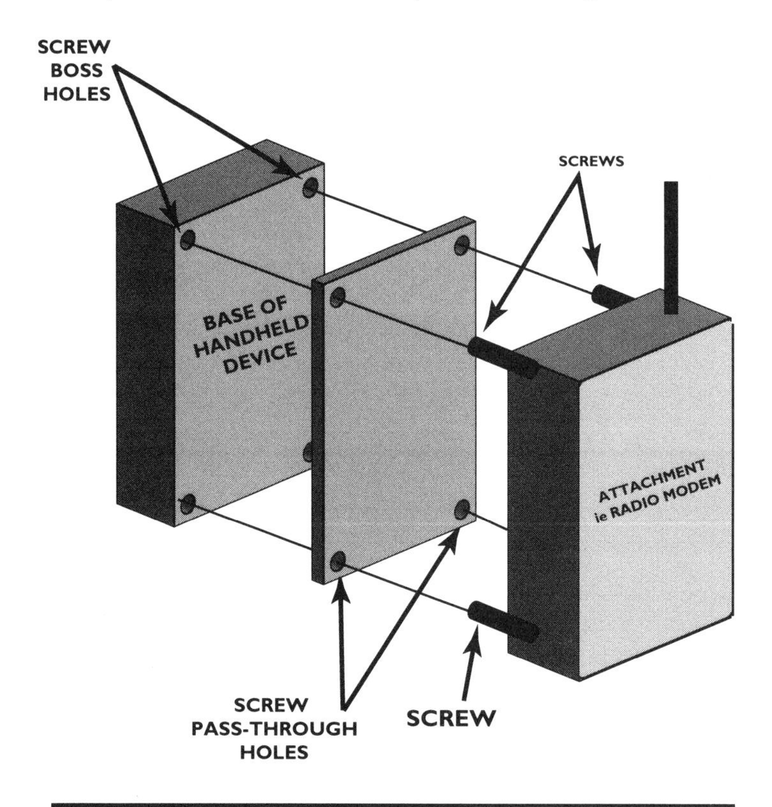

Figure 9.5 Using screwbosses for hardware integration.

Computing devices and peripherals can also be vehicle mounted to provide mobility. In addition to moving the burden of the device from the mobile consumer to the vehicle, additional benefits can be realized from vehicle mounts. The power to run the computing devices and peripherals can be supplied by the vehicle battery or another power source mounted in the vehicle. This allows much more power than can be carried by one person. The vehicle can also provide a location for the antenna to be permanently mounted in a fashion that increases its effectiveness in transmitting and receiving mobile data. Vehicle mounts often give the mobile consumer the flexibility to use the computing device both in and out of the car or truck. As such, it provides a platform for operating the mobile system as well as a resting place or temporary storage location for the portable equipment. Many mounts allow the device to be slid in and out of a tray or holder. The mount simultaneously connects the device to a power source and a communications interface when it is slid into the tray or holder. The mobile computing device can then recharge and transmit mobile data while the consumer is in transit. Vehicle mounts are a popular alternative for wireless computing systems deployed to fleet-based mobile consumers. (see Figure 9.6).

Figure 9.6 A typical vehicle mount.

Carrying cases are a more popular and less expensive means of combining physical components into a single unit. They are usually designed to be hand carried, worn on the shoulder (i.e., in a harness or holster) or worn on a belt (Figure 9.7).

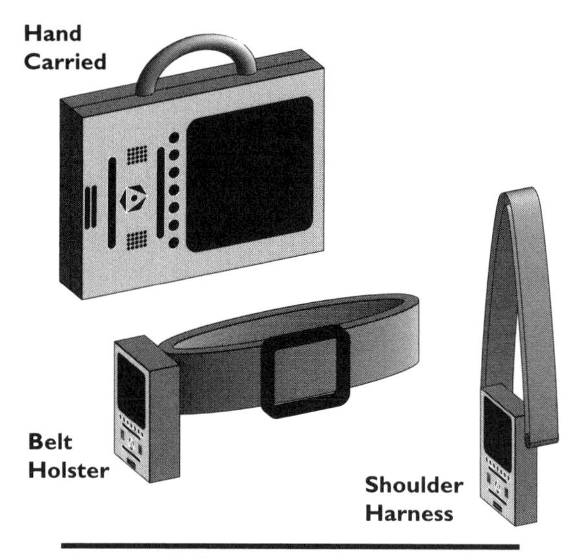

Figure 9.7 Three carrying case styles

Regardless of the manner in which the case is carried, the design of the case must allow portability as well as operation. Certain factors which affect the way the device is operated must be accounted for. First, the user interface must be easily accessible. Buttons or keypads on handheld devices and screens on pen computers should be accessible without having to open cases that are worn. Keep in mind that pen computers with pressure-sensitive screens, or resistive film digitizers, can usually be used when lightweight plastic covers the screen. Pen computers whose screens operate with magnets activated by electric (screens with an electromagnetic digitizer) require that the stylus make direct contact with the screen. The carrying case, therefore, must allow easy, unobstructed access to these screens.

The mobile consumer should be able to reach keyboards on notebook or palmtop computers with ease. When antennas are involved, the case design should allow it to protrude and be pointed in different directions to improve reception and transmission. It is convenient for the mobile consumer to be able to use ports (e.g., IR, serial, TTY) and AC adaptors without having to remove the device from the case.

A challenge in designing cases is providing these operational features while constructing a case that is durable and easily carried or worn. Cases should be made so that corners, angles, and openings are reinforced to avoid being torn or worn out too quickly from frequent use. Carrying cases can provide an extra measure of protection against the elements. Materials used in constructing the case can help protect against rain, show, heat, and cold. They can also provide an added cushion in the event the portable device is dropped.

Whether you choose to construct a single device, use a vehicle mount, or tote a carrying case, attention should be given to making the mobile computing device easy to carry. The design and selection of portability options should be based on the routine and environment of the mobile consumer. They should increase the operability of the mobile device as well as its durability.

Developing a Wireless Computing System

Putting It All Together

Once the general design specifications have been finalized and the system components have been selected, development of the wireless or mobile computing system can begin. Since most systems will consist of separate components that must be used together, development and testing should begin by adding and testing each component, one at a time. This will make it easier to detect and pinpoint any problems that may exist between any two components. Whether you are purchasing off-the-shelf components for a horizontal system or have decided to develop software or modify hardware for a vertical system, this step by step testing procedure can help in identifying potential problems.

Testing Your Wireless or Mobile Computing System

1. Check each component separately.
2. Load and test all systems software.

3. Load and test all applications software.
4. Connect and test each peripheral, one at a time.
5. If you are using a modem or PC card, test them first.
6. Use the assembled system in the actual environment(s).
7. Check and test backup and support sources.

If a horizontal system is to be implemented, the components selected should work together and be easily assembled. It is wise to check with the vendor of each component before purchasing to ensure that they are compatible with each other. Most manufacturers and developers have certification labs which test their products with others to determine the level of compatibility. In some cases your system may require an extra piece of software or an adaptor for smooth operation. Whenever possible purchase components within a close timeframe and from supply sources that offer a money-back guarantee. Make sure that you have complete documentation and the telephone number for technical support. This will provide you with the time and resources necessary to test all components and ensure that they work together as planned. Problems can sometimes be rectified with a minor repair or by exchanging components for a different model or version.

If a vertical system is to be implemented, development should begin with the software. The project requirements and system design documents (see Chapters 1 and 2) will assist the software developer in identifying the required functions. The software developer must remain mindful of the hardware and communications environments. The user interface provided by the operating system should be exploited to maximize ease of use while minimizing consumption of hardware resources. Software functions should be written or coded efficiently to minimize and group I/O and processing. This will help to extend a precious commodity in mobile computing—battery life.

Software should keep the ever-moving mobile consumer aware of what's going on. When a response will take more than 5 seconds to be displayed on the screen, an advisory message should be given. The software should preserve as much information as possible in the event of battery failure. Since all operating systems are not power aware, and the mobile consumer may not monitor the remaining battery life, the software should help protect the mobile data. Each record or valid group of information should be placed in permanent storage as it is created or received.

Make sure that all integration issues are addressed as the wireless or mobile computing solution is assembled and tested. Maintain an open line of communication with all vendors and other support sources. Involve at least one mobile consumer in the preliminary testing of the assembled system. This will assist in identifying areas needing change and making adjustments prior to the pilot.

Conducting Pilots

The pilot is one of the most important steps in developing a wireless or mobile computing system. It is the live feasibility study. The pilot allows you to whether the intended benefits will indeed be realized. It provides a validity check of the time and resources necessary to successfully implement and rollout a wireless or mobile computing system.

For an individual, the pilot is seeing a demonstration or actually using the software. It is seeing or testing the notebook or palmtop computer or the personal digital assistant. A pilot for the individual is attending a trade show or visiting a computer store to try the software and hardware. It is giving careful thought to how and when you will use the mobile system in your daily routine. The pilot for the individual is doing everything possible to gain confidence in the use and value to be gained from a wireless or mobile computing system.

The pilot for a group of mobile consumers is a mini-rollout of the planned system. It helps determine whether the investment of resources will be for the good of the business. The pilot also provides information and direction on how to successfully implement and roll out systems.

A pilot should be designed to meet several objectives. First and foremost, it must determine if the system will perform as designed. It answers the questions—will it work? The pilot should focus on whether or not the business functions that the wireless or mobile computing solution has promised can be delivered. That means making sure that all of the software functions operate as intended. It also means making sure that the hardware can be operated as planned. During the pilot phase you must be sure that the mobile consumer is able to use the system as it was intended. Are they able to easily integrate it into their daily routine and perform all of the functions of the system? Are they able to operate it, carry it around, and send information where they need it to go? If the system is a wireless computing system, the pilot determines if information can be transmitted. Will it transmit in all locations and all environments? Will mobile consumers who are inside buildings be prohibited from gaining wireless access when they need it? If access is not always available, the pilot allows you to determine the impact of temporary non-functionality on business operations.

The support plans and resources which have been planned and put in place should also be tested during the pilot. If the mobile consumer needs to replace a unit, the pilot will verify how long it will take to get that unit replaced. It provides the opportunity to test the established backup systems and procedures. If a hotline is part of the support plan, it should be operational during the pilot. Does the mobile consumer understand how to use it? Are they able to get through and access it easily during the course of their workday? The training methods should also be tested. It is essential that the training be effective. The mobile consumer must be able to understand and absorb the routines for using the wireless or mobile computing systems within a reasonable timeframe. The pilot allows you to determine how much time must be taken out of their day to learn how to use this system. It also clarifies the ramp-up speed and time which is required to make the system work. Essentially, the main function and first objective of conducting a pilot is to make sure that the system will work. That means that the mobile consumer can learn, understand, and actually operate the system effectively as they perform their daily routine.

The pilot also provides an opportunity to identify and assess the cultural impact that the wireless computing system will have on the mobile consumer or the field workers of the company. Cultural effects can include changing the manner in which the mobile consumer delivers a product or service to the customer. They can include altering the way in which different departments work with each other. The system should not intrude on the manner in which operations are performed. It should complement operations. A mobile system should be implemented without disrupting the flow of the business. It should not impede the mobile consumer in completing primary functions. The goal of the mobile consumer is not to use the wireless computing system, but rather to perform a certain task or operation. The purpose of the wireless computing system is to enhance the ability of the mobile consumer to perform that task and to enhance the value of the information that might be collected or disseminated during the performance of that task. The system enables the production of mobile data. Mobile data should be of value to the mobile consumer as well as to the corporation or business entity. To that end, it is very important to understand the affect, particularly the cultural impact, that a wireless or mobile computing system might have on one of the most valuable assets of the business—the employees.

If the mobile computing system is presented properly, the mobile consumer will view it as a real asset. If the mobile computing system is not presented properly, in many cases, the field worker or mobile consumer might view it as a threat. He or she might show a reduction, rather than an increase, in productivity. If the wireless or mobile computing system is poorly presented it could also cause tension or dissent among the field force and management. It could be viewed as a threat to their job or position and, therefore, cause more disharmony than good. The pilot can help to identify any ill feelings which could arise during the rollout of the system. This will allow the systems planner to prepare to avoid such problems in the final design or rollout stage. Those changes may not always have to do with the actual software, hardware, or components of the mobile or wireless system. In many cases those changes may be in how the system is presented to the mobile consumers it is intended to serve.

Another objective in conducting a pilot is to identify the ease with which the mobile consumer is able to use the system. In particular, the ergonomic factors must be taken into consideration. Is the mobile device easy to use or is it a burden? The pilot can show whether the portable device is easy to carry throughout the day. It must also be determined whether the user interface allows the mobile consumer to easily use the system as they perform their daily tasks. The pilot may identify small changes or major changes that may be needed to facilitate use of the mobile device.

And last but not least, another major objective of the pilot is to determine whether or not the system is worth it. The financial measures for conducting and using the system must be tracked during the pilot. The business case not only outlines the intended results or benefits from the system, it will also outline the that which should be used to determine whether or not these objectives have been achieved. At the outset of the pilot, begin the tabulation of all costs involved in implementing and rolling out the system. Tangible and intangible benefits should be identified and tracked. Returns should be calculated. Determine the payback period. How long will it take to actually reap the benefits of the system, so that it will pay for itself? Will it pay for itself in a month, or will it take years? Determine how long it will take to reap the benefits and have them compensate for the costs, both out-of-pocket and intangible, of implementing your system. Spend time identifying and measuring all costs and all benefits that have been expended or realized during the course of the pilot. This is critical in assessing the financial viability of the system.

The objectives of the pilot should be carefully defined and documented so that they can be viewed, analyzed, and measured as the pilot is being conducted. The pilot objectives should remain clear for they provide the framework and the basis for not only ascertaining if the pilot has been successful but also in determining what should be done to improve the likelihood of successful implementation of the wireless or mobile computing system in question.

In addition to providing well defined objectives, several other items should be considered and planned for when preparing for the pilot of a wireless or mobile computing system. First, keep in mind that the purpose of the pilot is not to demonstrate how technically complex or efficient the system is. Rather the purpose is to demonstrate how easily this tool can be used to enhance the value of mobile data to the business enterprise. Therefore, one of the things to keep in mind during the pilot is to keep it simple. The system being piloted, as well as the manner in which it is presented, should be clear and straightforward.

It is necessary to have a comprehensive plan for the pilot. Make sure everyone knows what's being done and their role in the process. All members of the project team should be involved in the planning of the pilot. Every member can contribute to ensuring that the pilot objectives are properly set and so that the pilot can be conducted smoothly. The technical staff can identify any areas that must be kept in mind while using the equipment. The mobile consumer can identify any operational areas, procedures, or routine tasks that need to be adjusted or taken into account in the training or preparation. Management can help identify business objectives that must be kept in mind while conducting the pilot. Remember to keep all of the staff involved at the beginning of the pilot and when reviewing the results of the pilot. They play a key role in helping to determine if the pilot met the objectives originally set for the system.

Since the development and testing of wireless and mobile computing systems involves the coordination of several different parties, it is very important prepare a thorough and detailed plan prior to the pilot process. This will help you to determine every single step which must be taken to successfully assemble, design, and construct an end-to-end system. Know what equipment must be delivered and what the lead times are for delivery of that equipment. It is important to identify what is necessary to provide the host interface and what steps have to be taken to make that interface work. It is also important to know the schedule of the participants in the mobile pilot so that everything can flow smoothly, the ill effects on business operations while conducting the pilot are minimized (Figure 10.1).

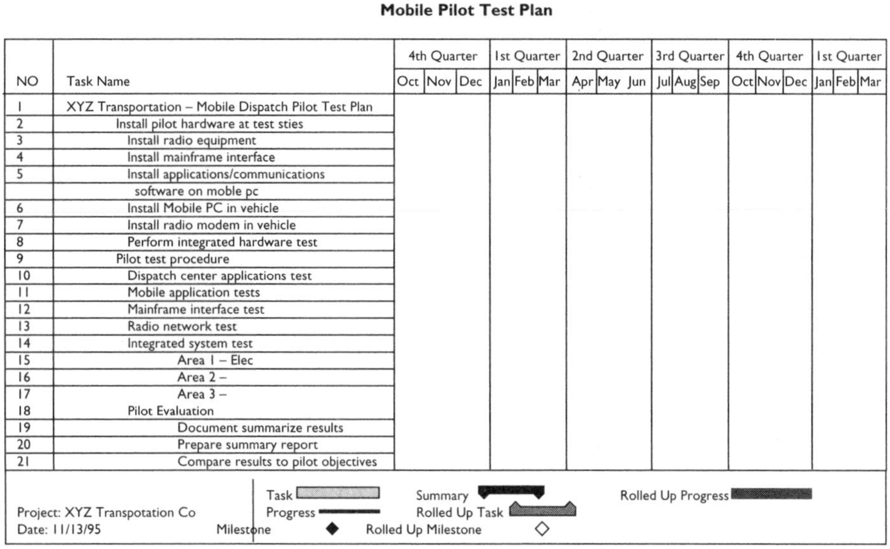

Mobile Pilot Test Plan

		4th Quarter			1st Quarter			2nd Quarter			3rd Quarter			4th Quarter			1st Quarter		
NO	Task Name	Oct	Nov	Dec	Jan	Feb	Mar	Apr	May	Jun	Jul	Aug	Sep	Oct	Nov	Dec	Jan	Feb	Mar
1	XYZ Transportation – Mobile Dispatch Pilot Test Plan																		
2	Install pilot hardware at test sties																		
3	Install radio equipment																		
4	Install mainframe interface																		
5	Install applications/communications																		
	software on moble pc																		
6	Install Mobile PC in vehicle																		
7	Install radio modem in vehicle																		
8	Perform integrated hardware test																		
9	Pilot test procedure																		
10	Dispatch center applications test																		
11	Mobile application tests																		
12	Mainframe interface test																		
13	Radio network test																		
14	Integrated system test																		
15	Area 1 – Elec																		
16	Area 2 –																		
17	Area 3 –																		
18	Pilot Evaluation																		
19	Document summarize results																		
20	Prepare summary report																		
21	Compare results to pilot objectives																		

Project: XYZ Transpotation Co
Date: 11/13/95

Task · Progress · Milestone ◆ · Summary · Rolled Up Task · Rolled Up Milestone ◇ · Rolled Up Progress

Figure 10.1 A mobile pilot chart.

It is important during the pilot to maintain support from several different groups. That includes the mobile consumers. This is essential, for the group that is actually going to be using the system should support it. Attention should be given to customers who might be involved in, or benefit from, the mobile system. Their feedback should be solicited. Management needs to be kept abreast of the pilot. This can be very important because in many cases the pilot requires personnel to be pulled from their daily routines. This may impact certain business operations. Also additional funds might be needed prior to the conclusion of the pilot, so management must remain involved throughout. Whenever possible, vendors should be involved. These are the companies supplying the products to make the wireless computing system a reality. The vendors have a vested interest in the success of the pilot, therefore, the lines of communications should be kept open so that the vendors who are involved in the project can provide support when it's needed.

An essential step in the pilot project is to manage expectations. The expectations of the mobile consumers, of management, and all parties involved need to be managed. Many people expect phenomenal results from new technologies. These results cannot be achieved until personnel are trained and routines have been altered. This will not happen overnight, however extraordinary the products might be. A proper perspective must be maintained of exactly what people should achieve as a result of using the system. It is very important to manage expectations throughout the pilot to be sure that people do not perceive that they are missing results that they were never intended to receive. The proper level of expectation will also help ensure the successful development and full rollout of the wireless or mobile computing system.

As the pilot is being conducted, several things must be taken into consideration. We must pay close attention to how the system is operating and be aware of the functionality of the system. We must also be mindful of the acceptance of the mobile consumer. In rolling out a mobile system to a large field force we want to keep an eye on how well they are accepting the system so that we can present it to them properly upon rollout. Many wireless and mobile computing systems not only affect the immediate consumer or worker group that is using it, the information it sends or gathers and the operations it performs will affect other areas of the enterprise. During the pilot phase, be aware of the scope of the operation and the range of the effects of the system. Understand who will be affected, where those people are located in relation to the corporation or business, also in relation to the way the business affects the environment in which it operates. Taking care and taking note of this can help to avoid problems that may arise down the road as a result of using a wireless or mobile computing system.

At the end of the pilot several steps must be taken to assess the results. The project manager or pilot conductor should make sure that all the findings and results of the pilot have been carefully documented. Once the results are documented and reflect the plans laid out before the pilot began, the appropriate people from the project team should analyze the results. Careful analysis of the results of the pilot allows you to determine what issues must be addressed before the final system is produced and rolled out. The analysis of the results also allows you to determine what additional steps need to be taken to ensure the successful rollout of the system. Additional steps might include revised training, different notification procedures, and identifying which groups or subgroups should be rolled out and in what order, based on the combined effect that these groups have on the business operations. The results of the pilot need to be viewed and analyzed very carefully because this is a key step in determining what has to be done to ensure the successful final design, production, and rollout of your system.

Once all of the results of the pilot have been reviewed and analyzed they should be measured against the initial business case. The reason for this is obvious. You need to determine whether or not the system should be rolled out. The results from the pilot might provide information to allow the business analysts to adjust projections in the business case. The pilot might reveal that the intended benefits will not be achieved as initially expected because of unforeseen effects that the system has on business operations. Or, the pilot might identify additional costs that must be incurred to successfully implement the system. Likewise, the pilot might identify areas for potential savings by successfully implementing and rolling out the system. Careful comparison of the pilot results with the business case will allow management to determine whether or not the system should be implemented.

If the results support final design, production, and rollout of the system, they should be used to help enhance and finalize the design of the system as well as lay out the implementation plan. These are other key benefits that the pilot will provide. It will identify all key steps that are necessary to ensure success of the development and implementation of a wireless or mobile computing system. Remember the pilot is essential. A successful pilot not only allows you to streamline the development process, it also allows you to effectively implement your system. A successful pilot will identify areas where you can contain costs and help you to avoid unnecessary expenses as you take on rollout of a much larger system. Even if a pilot is not successful, it's still essential because the process has prevented you from wasting money attempting to design, develop, roll out, and fully deploy a system which would not have yielded the expected benefits.

The pilot is one of the most important steps in the design, development, and implementation of a wireless or mobile computing system. It should be carefully planned and objectives should be well documented. The right staff should be involved throughout the pilot process. Care should be taken to document all findings and results to allow the planning and operations staff to ensure the successful system deployment.

Implementing a Wireless Computing System

Installing the Components

Assembly and Preparation

The implementation of a wireless or mobile computing system is paramount to the success of that system. The implementation involves every single step that it takes to get a completed, designed, and developed system from the drawing board, the point at which it's assembled in prototype form, into the hand of the mobile consumer, who will be using it. Therefore, implementation of the system involves delivering all of the components, making sure that all the support is available, and training the mobile consumer to properly and successfully use the system. Implementation of a wireless or mobile computing solution is more a job of logistics, training and human intervention, than it is a job of understanding how the system works. Remember that a wireless computing system is a tool that's used to enhance the daily operations of a company. Moreover, a wireless computing system is a tool to help the mobile consumer in their daily routine. Therefore, to successfully implement a new system you have to smoothly meld the use of that system into the operations of the company and the mobile consumer. Invariably, this is going to cause some temporary lack of productivity or inability to complete the daily routine.

You're introducing something that has not been in daily. And the item or tool that you're introducing must be thoroughly integrated with the operations of the entity, for it has a big role in the successful transmission of information from the mobile worker to the host computer. Therefore, as you look at implementation of the system it's important to consider everything that's being done in the day-to-day operations and how the daily routine will be impacted by the system.

The business case will help with that process but it's important to understand how the system will fit into the day-to-day operations so that the components of the system can be installed in the most efficient manner. It is critical to have a detailed step-by-step implementation plan (see Figure 11.1).

Mobile Implementation Plan

		1st Quarter			2nd Quarter			3rd Quarter			4th Quarter			1st Quarter			2nd Quarter		
NO	Task Name	Jan	Feb	Mar	Apr	May	Jun	Jul	Aug	Sep	Oct	Nov	Dec	Jan	Feb	Mar	Apr	May	Jun
1	XYZ Transportation – Mobile Implementation Plan																		
2	Final System Testing																		
3	Test radio equipment																		
4	Test mainframe interface																		
5	Test applications/communications software on moble pc																		
6	Dispatch center applications test																		
7	Mobile application tests																		
8	Integrated system test																		
9	Hardware procurement, delivery, installation																		
10	Witness factory acceptance tests																		
11	Delivery of mobile pcs																		
12	Delivery of pc mounts																		
13	Delivery of radio modems																		
14	Install PCs, modems, mounts																		
15	System Delivery and Installation																		
16	Area 1 Installation																		
17	User training – Area 1																		
18	System site tests – Area 1																		
19	System site acceptance – Area 1																		
20	System operational – Area 1																		
21	Area 2 Installation																		
22	User training – Area 2																		

Project: XYZ Transportation Co – Mo
Date: 11/13/95

Task ▒ Summary Rolled Up Progress
Progress — Rolled Up Task
Milestone ◆ Rolled Up Milestone ◇

Figure 11.1 Implementation plan.

This plan allows you to effectively coordinate all of the parties who will be involved in the implementation of the system. It will also allow you to install the components so as to cause the least amount of disruption in the daily operations of the business.

As you look at installing the components, there are several different pieces which must be put into place first to make sure that the effects on the business operation are minimal. First of all, in the pilot and testing phase you should have identified everything that worked together, and by going through that process you should be able to identify exactly what must be put in place and when to allow the mobile consumer to pick up the portable unit and begin using it. In other words, you should know what host system software and functions have to be put in place. As you begin looking at the host system, in many cases, there are items that can be put in place without disrupting the operations, changing the way business is performed, or the way the host system is run *until it's time to implement mobile systems*. Prior to rollout, system operations and load tests should be performed with the host system. Everything that is necessary to make the wireless or mobile computing system function, should be in place and ready for operation. This includes each and every component necessary to support all mobile consumers. The components which must be installed include software and hardware at the mobile and host sites. The hardware necessary to handle large numbers of mobile consumers may include multiplexors, front end processors, and modem pools. Installing the components for a wireless or mobile computing system first means having those components available to assemble them into the system that has been designed.

Specifically for a portable device, you need all the necessary components: the hardware device, the software, if it's a wireless system you need a radio modem or PC card, carrying cases—any shoulder harnesses or portable bags that must be used. The actual mobile or portable device that the mobile consumer will be using should be designed, developed, and readily understood as a result of the pilot. Therefore, as you look at installing those components you must make sure that you've placed the orders with all of the vendors to deliver the volumes of products that you will need in order to assemble the system and deliver it to the mobile consumer. Keep in mind that many devices and many components, whether it's hardware, software, or network items, when delivered in large quantities need to be ordered with considerable lead time to ensure that you get exactly what you want, configured as you need it. Secondly, you need to allow enough time to put all the pieces together or actually set up that portable device for the consumer.

Again if you're looking at large quantities, allow enough time after receipt of the items to install the software, test the hardware, and, if you have a custom solution, allow sufficient time to do any of the customization that may be necessary and make sure that the final device works as planned. If you are implementing a wireless computing system you need to make sure that all of the subscriptions to the network for each mobile consumer are in place and have been identified and are working with each item. In many cases an individual portable unit will be given to a particular person and that person must be able to identify and track that unit. In other cases, the unit might change day to day, for different people, and you need to prepare for that as well.

Keep that in mind as you set up subscriptions on the wireless network. If you will be using a docking station, or some other equipment to recharge the unit and perform volume data transfer over landline communications during the evening, it is important to have all of that equipment installed and in place so that the mobile consumer will have everything that's needed to use the system fully. Another step in installing the components is making sure that support is in place and ready to go. If you're using a hotline to answer questions or provide direction for the mobile consumer, that hotline number should be given to all and it should also be adequately staffed and ready to go.

If you are using spare parts or extra units to provide spares in the event that any of the portable devices fail to function properly, those spare parts and units should be configured, installed, and ready to go. That will allow you to quickly get another device to a mobile consumer who has lost the functionality of their device due to breakage, loss, or theft.

Your implementation plan should serve as a guide to everything that's necessary to deliver a completed system to those who will be using it. That guide will lay out everything that needs to be done, from ordering the equipment to making sure that every component is in place and ready for the mobile consumer to use it. Accurate and timely installation of each and every component of the wireless or mobile computing system is essential to the successful launch and use of that system.

Deployment

The second phase in implementing a wireless or mobile computing system is preparing for the rollout. Preparing for the rollout includes a number of things. You have to advise and notify everyone in every location who will be involved or impacted by the implementation of a mobile system. A very critical step in rolling out the system is recognizing the number of mobile consumers who can be outfitted with the system without substantially disrupting the operations of the business.

If you are implementing a system to a small group of employees, i.e., 25 or less, it may not be as critical to spread out the timing of delivering that system to those people. Depending upon their daily operations, they might be able to take all their systems at one time and use them without substantially interrupting business. In most cases, the mobile consumer is a crucial element to the flow and operation of the business. So, when you are disrupting their routine and adding something new, you have to plan and allow enough time for the business to run without them being at full productivity. For that reason, it is strongly recommended that staged rollouts take place when implementing wireless or mobile computing systems. That is, you provide the system to a limited group of mobile workers at a time.

Again, depending on the effect that rollout has on the daily operations of the business, as well as on the volume of people that you have to rollout a system to, you need to map out a schedule that indicates how many people will receive a portable unit and when each person will receive a unit (see Figure 11.2). In other words, not only the date that those people will receive the units but how much time will transpire between each group of mobile workers receiving their portable devices and actually coming up to speed and using them. The number of workers who will receive portable systems (and be included and integrated into the full functioning system) depends on the complexity of the operations, the importance that the worker or team of workers has on maintaining the daily operations of the business, and also the total number of workers that you have to roll out to. Another consideration is the difficulty of maintaining the operations of the business without *everyone* having a portable device once the system is in place.

SAMPLE ROLLOUT PLAN

Implementation Step	Completion Date
1. Delivery of System	
2. Training	
A. Staff Consumers	
B. Mobile Consumer	
3. Pilot	
4. Training Mobile Consumers	
5. ROLLOUT: Delivery Stage 1 Units	
6. Training Mobile Consumers	
7. ROLLOUT: Delivery Stage 2 Units	
8. Training Mobile Consumers	
9. ROLLOUT: Delivery Stage XX Units	
10. Run Parallel with Existing Process	
11. Exclusive Use of the Wireless Computing Systems	

Figure 11.2 Sample rollout plan.

Staged rollouts should include 10 to 50 mobile workers at a time. The complexity of the system is another item that determines how quickly a staged rollout takes place. Again, you're looking at anywhere from 10 to 50 mobile workers who would be getting this system and using it at a time. Generally no more than 10–20% of a workforce will take on the system at one time. This will allow you enough time to make sure that all of those workers have everything that they need to use the system and have been thoroughly trained on exactly how it will work—and that includes not only how to operate it on a daily basis, but how to care for the system when they have completed using it for the day. That care may be putting it in a docking station or taking it home and recharging it and uploading information.

Bringing the mobile worker up to speed on the system also involves what it takes to replace the system or get help and assistance in the event that something goes wrong. They not only need to know what to do to care for the system, they also need to know how they're going to perform their operations without having this new tool, or mobile computing device, that will be temporarily unavailable. Will they revert to pen and paper? What measures have been put in place to allow them to continue operating *without* the new system once it becomes a staple in their daily routine? So there are several things that have to take place before you can consider the mobile worker proficient, or ready to roll, on a wireless or mobile computing system.

Many of these factors will have been outlined and identified during the pilot, which will help you to plan a rollout schedule that will not only give the mobile worker adequate time to prepare and become comfortable in using the system, it will also give the business operation sufficient time to absorb the new system into the enterprise. It is essential, particularly in systems with large numbers of mobile consumers, that a staged rollout plan be in place. This plan will allow you to understand who has to receive what equipment and when. It is an important step in implementing your wireless or mobile computing system with the least amount of disruption.

Training the Mobile User

The most important step in successfully implementing a wireless or mobile computing system is training the mobile consumer. Again, the mobile consumer is the person or persons who will determine the true success of the wireless or mobile computing system. And if they are not able to use the system to achieve the objectives that have been set, the system will be a failure. It is critical, therefore, that the mobile consumer be properly trained. The training will not only involve showing them how to operate this new, far-reaching device, the training, more importantly, involves helping them to understand how this new tool will help them as they go through their daily operations. With many companies, particularly those whose mobile users were not properly involved and did not buy into the mobile system, rollouts have been delayed and results not achieved as expected. Although the mobile system proved beneficial and more than paid for itself, the benefits were limited by mobile consumers who did not produce as much as they should have been able to with the system.

The training also must include ways to sell the mobile consumer on using the system. The mobile consumer, in other words, must want this to happen and want this to work, if they are to put their best efforts forward in using the system as efficiently as possible. Many companies that have large mobile workforces will actually use one or more of the members of that workforce to train others. Therefore, one step in preparing to train the mobile worker is to prepare a training class for the trainer. The member of the mobile workforce who will be training others must not only be taught how to use the system but also how to demonstrate the system and train others to use it.

Training the mobile consumer cannot take place simply in the classroom. In most cases the systems are set up to be simple enough—and this should be the case with all systems—that there's never a classroom component to training. The mobile consumer must understand how to operate the mobile device, how to carry the device, how to protect it against the elements of rain, sun, wind. If they're carrying other items in addition to the device it is helpful to show how they can carry everything together with ease. Many package delivery carriers have found that they simply put their device on top of all of the boxes that they carry, or else they wear a special harness for the device.

Even things that may be inevitable and intuitive to the consumer should be made as easy as possible for them. So the training should show the mobile consumer everything that's necessary and everything that's helpful to use the device in their day to day work. Not only should the training take place in the real work environment—that means along the truck routes, in the warehouses, in the shipyards, or in front of the customers as you're collecting survey information, the system should also be used immediately prior to the actual continual use by the mobile consumer. In other words, it is not wise to train the mobile consumer one day and have them start using the system at a later time. There has to be hands-on, immediate rollout of the system. People tend to learn to use systems, particularly computing systems, *as* they use them. That learning is reinforced by continual and repetitive use. Therefore, going through training at one point and having too much of a time lapse before the system begins being used usually makes that training virtually worthless. Many people will not remember the fine points of how to use the system. It is important, then, to conduct the training as the mobile consumer is beginning to use the system. Many companies and individuals have found that some sort of small pocket-sized guide is very helpful. The guide should be very easy to read, contain pictures whenever possible, and show just the functional points of how to use the system. This will serve as a tickler, or reminder, for the mobile consumer on how to perform certain functions, less cumbersome than a detailed training example or guide on how to use the system. The guide can be carried with the mobile consumer and give them a measure of comfort in being able to look up how to do certain things, or perform certain functions.

Training the mobile consumer not only means showing them how to operate the system in their day-to-day routine, it also means showing them what to do in the event of a problem. The problem may include the system going down because of battery failure, or damage to some of the components, like a bar wand or stylus. The problem could be having lost the case, a component, or the actual portable device itself. Helping the mobile consumer understand what to do in the event of problems generally means demonstrating how to use the support forces that have been put in place. This might mean making sure that replacement equipment is provided in the event that something is broken. It might mean actually going through that process—doing it—so that the mobile consumer will see what it takes to get a new unit up and started. It could also mean calling a hotline number, if you have one set up, so that the mobile consumer will go through the process of making the phone call and understand what to do. It is not only important to train the mobile consumer to give the right information to the person on the hotline, it's also important for the consumer to understand that there is another level of security or assistance in their use of the mobile system. They will know that they are not being left out to dry and that there are options in place in the event that the device does not work.

Training the mobile consumer is the final step in successfully implementing a wireless or mobile computing system. This step cannot take place until all of the components have been installed and the rollout has begun. This step will not be successful in the deployment of a large system unless you rollout in stages. And, finally, this step is essential to make sure that the operations of the business will continue without major disruption

Maintenance and Support

Elements of Support

Wireless computing systems often become an integral part of a company's operations or an individual's daily routine. Without the proper maintenance and support, failure of the system may occur causing major disruption to operations or one's daily routine. Disruption of operations could produce substantial losses for a company. Impairing one's daily accomplishments can have cascading or far-reaching effects. It is imperative that support requirements be thoroughly reviewed, carefully planned, and set into motion at the point of initial implementation.

To provide complete support for a wireless computing system, assistance must be available at every step along the flow of mobile data. That is, support should be available from the time mobile data is sent, received, or collected to the time data mobility is no longer necessary.

There are five areas of support that must be considered when planning support for a wireless computing system:

1. Training the Mobile Consumer
2. Providing Comprehensive Hotline Support

3. Portable Hardware Maintenance and Repair
4. Facilitating Software Upgrades on the Mobile Device
5. Allowing for Network Outages

Training the Mobile Consumer

The mobile consumer can make or break the success of a wireless computing system. Some managers may believe they can exert their authority and force the mobile worker to use the system. This method can prove disastrous. Many mobile workers are union or blue collar employees who may view this new intrusion as a threat. Others may not be convinced of the real benefits to be gained. A less-than-enthusiastic level of participation will often lead to less-than-desirable results from the system. These ill results are often intensified when representative(s) of the mobile force are not included in the system planning and design process. In short, "training" the mobile consumer begins with selling the idea of using a mobile computing system. This sales process should begin long before rollout. It should begin early in the planning process.

Once the mobile consumer group has bought into the plan to implement a wireless computing system, it becomes a partner in the training process. Overcoming this hurdle makes a difficult job a little easier. Unlike current popular computer training courses, however, training for the mobile consumer must allow for the special circumstances and requirements that mobility creates. For example, he or she must understand what tasks can be performed "while moving" and which should wait until he or she is stationary.

The mobile person may need to learn how to best carry the portable device and how to care for it in inclement weather. They may need to be able to recognize when new information has been received—by themselves or by the person to whom they sent information.

If the mobile consumer is part of a group that will begin using a mobile computing system, the training should be tailored to the needs of that group. For mobile consumers who seek "generic" training, the instruction should include sections on mobility issues (i.e., using software on the go, caring for portable hardware).

Although these issues are taken into consideration when preparing training for implementation of a wireless or mobile computing system, they also play a key role in maintaining these systems. New mobile consumers will likely join the mobile computing force. As updates and changes are made to the system, mobile consumers must be made aware of these changes; and in some cases, they will require new or refresher training courses. The training of mobile consumers is essential to providing ongoing maintenance and support. As long as the wireless or mobile computing system is in use, it should never end. It is imperative if the benefits gained from the system are to be continually realized.

Providing Comprehensive Hotline Support

When a mobile consumer experiences problems with his or her system, he or she must have a means of immediate support. Once the wireless or mobile computing system has become an integral part of the routine, the mobile consumer is likely to exhaust all possibilities before resorting to pen and paper or another nonautomated tool. It is impossible to prescribe solutions for every potential problem in the documentation provided to mobile consumers. Although a telephone may not be readily available, one is seldom terribly difficult to find. Rather than have the mobile consumer depend completely on written guides or portable on-line documentation, it is often advisable to provide telephone hotline support.

To provide an adequate level of support, a hotline must be easily accessible and adequately staffed with competent personnel. Accessibility means that the mobile consumer must be able to dial in from most remote locations. This often requires that the mobile consumer have a calling card or that the hotline have an 800 number. Adequate staffing means having enough people to answer critical questions during peak periods.

The hotline staff must have comprehensive training and complete support tools to enable the proper level of support. Not only should the hotline staff understand all components of the system (hardware, software, and network), they should also be able to recognize commonly recurring problems. This can be attained by training as well as a robust support database system that is continually updated. Another support tool that should be available to the hotline staff is direct access to high-level technical and operations support. This will allow the hotline staff to expedite extended support when necessary. High-level technical support should include system development and vendor personnel. Operations support should include those persons who manage field operations, those who manage spare parts or portable devices for the systems, and those who deliver training.

A hotline may not be warranted for all mobile computing systems. In many cases, the cost of providing a hotline far exceeds the benefits to be gained from the system. In other cases vendors may provide a hotline that is more than adequate for the type and magnitude of system developed. When a customized, fully integrated wireless mobile computing system has been rolled out to a field force of 1,000 of more, a hotline is often warranted.

Portable Hardware Maintenance and Repair

The support process should provide for the maintenance, repair, or replacement of all hardware in the wireless or mobile computing system, especially the portable computing device. The hardware in the network or host computer is usually maintained by the network carrier and the MIS department. The portable computing device, on the other hand, is generally a new acquisition, purchased to complete the system. It sustains the greatest amount of repeated physical use. A portable computing device is subject to wear and tear from several things, including rain, sun, humidity, and being dropped. To prolong the useful life of a portable device it must be properly maintained. Likewise, to avoid prolonged disruption of the benefits of wireless and mobile computing, it is imperative that damaged and broken units be replaced expeditiously.

Many hardware vendors offer extended service contracts. Most of the independent service companies who offer depot or on-site repair of personal computers, however, do not yet service all portable computing devices. This is particularly true if the portable computing unit/device is a custom configuration. Whether you are responsible for the deployment of one, one thousand, or ten thousand devices, it is advisable to make provisions for maintenance and repair. If you are using a small number of portable computing devices (i.e., 50 or fewer), it is usually more cost-effective to purchase a service contract with the purchase of the hardware. Many vendors charge more for a service contract after the warranty period has expired. Some even charge more if the contract is not purchased with the hardware. It is prudent, therefore, to take service requirements into account when planning for the system. In many cases it may be a major criterion in the selection of hardware.

The payback period achieved with some wireless and mobile computing systems is actually less than the warranty period. A system planner in a major pharmaceutical company once actually referred to their $1.5 million initial rollout as a "throw-away system." He realized that the benefits his company would realize were far greater than the cost of the initial rollout. Since he planned to maintain the software and hardware interface (i.e., future products purchased had to be identical or similar), the underlying software and equipment used in the initial rollout could be replaced with minimal effort. Before purchasing a service contract the system planner should evaluate the system design and review the length of the payback period. If the design allows the system to be easily adaptable to new platforms and the payback is less than the warranty, a service contract may not be necessary.

The replacement of computing devices in a mission-critical system can be satisfied by stocking a few extra devices. The time-value of information may mandate that a malfunctioning unit be replaced immediately—regardless of the number of devices deployed. The cost of providing new replacement units is much greater than the price of purchasing and storing or warehousing them. The additional cost of service, deployment, and tracking are not insignificant. Software may have to be loaded before shipping it to the mobile consumer. If the software has not been backed up, it may take extra time and effort to reconstruct what must be replaced. Peripherals may have to be attached and tested before shipping. Once the replacement unit is ready, it must be sent to the mobile consumer. Follow-up is necessary to ensure that the mobile consumer has received and is using the replacement device. Finally, all damaged and replacement units must be tracked and accounted for. To maximize the value of the warranty and minimize costs, a record must be maintained that shows when devices were purchased, their warranty period, when they were repaired or replaced, etc. In order to plan and account for the cost of servicing a portable computing device, it is necessary to identify and understand every step that is required to empower the mobile consumer.

Regardless of the time-value of the application, an individual or company is not advised to engage in self-service unless one or more of the following criteria have been met:

- a replacement unit is required before the hardware vendors can provide a replacement unit
- the cost of stocking and deploying replacement units is less than the service contract
- the required level of service cannot be secured or is not available

Should you decide to engage in self-service, make sure that you have a sufficient number of people on staff with the required skills. It is generally recommended that the number of extra portable computing devices equals 10–15% of the total amount deployed. While this percentage is based upon the mean time of failure and cost of servicing a typical portable computer, a lower percentage may sometimes suffice. The actual amount of spare units that are required depends on the durability of the hardware, the usage habits of the mobile consumer and the effectiveness of the training and support.

When planning service for a wireless computing system it is preferable to select a single source to service all components including hardware, software, networks, and training for the mobile consumer. If this is not possible remember that the portable computing device is often the most fragile component of the system. Make sure that is properly maintained and, when necessary, repaired or replaced in a timely manner.

Facilitating Software Upgrades on the Mobile Device

Software upgrades may be required every 6 to 12 months. If the system design does not include all critical functions or if bugs are encountered, upgrades will be needed more frequently. Not only should software upgrades be transparent to the mobile consumer, they would also be performed with minimal disruption of their daily routine. To ensure that these objectives are achieved, software upgrade requirements should be taken into account in the system design. Many wireless and mobile systems incorporate docking stations or other equipment to recharge the portable computing device and upload information to the host computer.

The information upload usually takes place at the end of the day and over landline communication networks, by which larger volumes of data can be transmitted more efficiently and usually at a lower cost. This process provides an excellent opportunity to perform software upgrades. The software necessary to allow the portable computing device to receive information should already be in place. The host computer can be programmed to send and execute the software upgrade when it receives a transmission from a device that has not been upgraded.

If the upgrade is not voluminous, or time critical, it can be performed over most wireless networks. The host computer can initiate and manage the software upgrade if it is connected to the wireless network. Public network carriers are also beginning to offer software upgrade services. If a software upgrade is performed over a wireless network it is advisable to notify the mobile consumer and gain his or her concurrence before the network-based upgrade is performed. The notification should include the time and duration of the upgrade as well as any steps that must be taken by the user to complete the upgrade. If the mobile consumer must re-boot the portable computing device, he or she should be reminded to do so. By giving the mobile consumer the opportunity to stop or delay a software upgrade, you avoid unnecessary disruptions to their daily work routine. This helps protect the value of the wireless computing system.

The mobile consumer can also perform software upgrades. Many are proficient in using computers, or will develop these skills after continual use of the wireless or mobile system. Software upgrades can be sent to the mobile consumer on diskette or PC card. This gives them the freedom and ease of performing the upgrade at their convenience. If the mobile consumer sends information to a host computer, the application can be programmed to notify the host of the current version of the software on their device. This makes it easy for the system administrator to track which software version is being used by every mobile consumer.

Whether you elect to use a landline network, wireless network, public network, diskette, or PC card, software upgrades should be transparent and painless for the mobile consumer. At all costs, avoid upgrades that require additional training. Upgrades should not be forced. The mobile consumer should be able to choose when they will be executed. Software upgrades are performed to enhance the functionality of a system. Avoid negating these enhancements by intruding upon the performance of the mobile consumer.

Allowing for Disruption

Once the wireless or mobile computing system has become an integral part of the mobile consumer's daily routine, what happens if it stops functioning? Think of how many times you could not schedule an appointment or purchase a ticket because the system was down. An individual's, or company's, inability to respond to customers needs will result in a temporary or permanent loss of business or, at minimum, frustration on the part of the customer. Although many customers are graciously accommodating, they often remember; too many problems or inconveniences will send them in search of your competitor.

The mobile consumer is a customer who is serviced by the wireless or mobile computing system. Provisions should be made to accommodate them if the system stops functioning. The malfunction may be a temporary network outage for a wireless computing system, or the unrecoverable damage or loss of the portable computing device. No matter what the cause of disruption, an alternative process should be in place which allows the mobile consumer to continue functioning.

This process should be developed during the system planning phase. The planning should involve key operations personnel, including mobile consumer representation, on the project planning team. The alternatives may include using a public or cellular phone, locating a fax, or simply resorting to pencil and paper. In most cases it is not feasible to cease operations altogether. To avoid this, as well as undue frustration on the part of the mobile consumer, remember to be prepared. Put an alternative process in place and prepare the mobile consumer to react in the event of a total disruption in the system. Preparing the mobile consumer should include reviewing this process during the initial training. While the development, implementation and maintenance of your system should strive to avoid and minimize system disruptions, they cannot be altogether avoided. Take steps, therefore, to minimize the impact on the mobile consumer or company, operations if disruptions occur.

Providing End-to-End Support

A wireless or mobile computing system usually requires the assembly or integration of products and services from multiple vendors and sources. If a problem arises, its cause is not always readily apparent. In addition to identifying and resolving problems expeditiously, it is important to have a congruous training and maintenance process. If a multi-product system is to perform harmoniously, it must be supported in the same fashion. Support is most valuable when it provides the mobile consumer with an added measure of security from the time that he or she begins using the system until their activities have been concluded for the day. Every facet of their use of the system must be provided for from diagnosing software, hardware, and network problems to the delivery of training hotline services, software upgrades, and replacement units. Ideally, this should be accomplished with a single point of contact that will provide the mobile consumer prompt and efficient service. This is the essence of end-to-end support.

End-to-end support is a multi-dimensional process. Is not simply ensuring the components along the flow of information. Indeed, it is a major accomplishment to keep a wireless computing system up and running—particularly when it is deployed to a large number of mobile consumers. Achieving this feat while engaging new technologies is worthy of congratulations and respect. It is not enough, however, to ensure the successful operation of many wireless computing systems. Successful operation means attaining the maximum possible benefits in the minimal amount of time. This cannot be achieved without empowering the mobile consumer with a realm of support that makes his or her use of the system effortless. Remember, focus of the mobile consumer must remain on performing a job or activity as they move about. This leaves little time to spend on operating or caring for a portable computing device. End-to-end support virtually eliminates the mobile consumer's need to be concerned about the wireless computing system itself. It allows the mobile consumer to use the system as a tool to accomplish their every moving objective.

Migration Issues

Preparing for New Requirements

The returns from most wireless and mobile computing systems will exceed their cost before the end of the useful life of most system components (i.e., hardware and software). Protecting the investment in equipment, therefore, should not be an overriding concern. As is the case with most computer systems, software and services comprise the majority of the expense of acquisition, development, and implementation. If the system is designed properly (in a modular fashion with popular protocols, etc.), the cost of adapting new technologies will be a fraction of the cost of the initial development. Given the limited resources of portable computing devices, most of the ongoing software enhancements are likely to be on the host. Since software and service costs are focused on empowering the mobile consumer in wireless and mobile computing systems, the investment that must be protected is that delivers those components which the mobile consumer interfaces with—the software and hardware interface.

The primary objective in migrating to a new wireless computing system is to avoid or minimize the need for additional training by the mobile consumer. If the hardware and software interface are maintained, the need for additional training will be minimal. The software interface is governed by the operating system and applications software. When migrating to a new system one should avoid making changes to existing screens, adding new screens, and changing the order of the screens. The hardware interface is determined by the portable computing device that was originally selected. Not only should the software run on any new hardware selected, the screens should look and feel the same. This means that the screen resolution should be the same or better. If a keyboard or buttons are used, the layout should not be appreciably different. When the portable hardware has a pen interface, the type of digitizer (i.e., resistive, electromagnetic) should not change. In short, whenever possible, the only changes to software or hardware interface that should be made are those which enhance usability *without* changing the look and feel.

Providing for Future Technologies

Migration Practices

When planning for the initial implementation of, or migration to, a wireless or mobile computing system, remember to provide for the future. This will help to avoid any pitfalls which may arise as you maintain and grow your system. Here are several migration factors which can serve as a guideline.

1. Adaptable environments
2. "Popular" protocols
3. Modular software
4. Scalable software and components
5. Plug and play components
6. Extended maintenance
7. Time-coordinated service and support agreements
8. Operationally compliant training
9. Dual component system
10. Disposal/reuse plan

Keeping Pace with Supported Components

Providing for future requirements is as much a part of system design and implementation as the development and integration of existing products and technologies. It is virtually impossible to predict every circumstance that may occur to impact your wireless computing system. History, however, tends to repeat itself. There are several practices that can be adhered to which will assist the systems planner in preparing for the future.

Systems upgrades are often warranted due to changes in:

- Business Requirements
- Consumer Demand
- Products and Services

Changes in business requirements often mandate a change in the definition or amount of mobile data and in the functions that the system performs. As mobile consumers become proficient in using the wireless computing system they will become increasingly aware of improvements that may assist their performance. New products and services hit the market at an electrifying rate. These will offer more functions, faster access, a greater array of options, more information, longer battery life, lighter weight, improved interfaces, and much more.

As these factors warrant consideration of a major upgrade or change to your wireless or mobile computing system, remember to look before you leap. Business requirements should obviously take top priority. Make a careful assessment of exactly what objectives must be achieved. Remember, *if it ain't broke don't fix it.* Requests from mobile consumers for changes may make a compelling case for migration. If their ideas would will result in a substantial improvement in business, the change is warranted. Finally, avoid succumbing to the fascination of new technologies. If the new products and services do not deliver features that serve business requirements, or those of mobile consumers, they are probably not needed.

Summary of Section II

Whether you purchase off-the-shelf components or elect to develop a customized system, the Project Planning and Life Cycle approach should serve as your guide (see Chapter 8). These are the steps to success for effective development and deployment of wireless or mobile computing systems. The system design requires careful attention, particularly for customized solutions. The design of a custom wireless or mobile computing system is a re-engineering effort which requires the coordination of all parties, skillful management, and fine-tuned integration. When performed properly the system design controls costs, promotes system integrity, and facilitates effective deployment. A wireless or mobile computing system is not complete, however, until it is used on a continual basis by the mobile consumer. This requires effective implementation, maintenance, and support. This book has presented the components and steps necessary to develop and implement a successful wireless computing system. The major points to remember are outlined on the next page.

Points to remember when planning a customized wireless computing system are:

1. Package the solution for the mobile user
2. Design for the future
3. Minimize the amount of data sent over the wireless network
4. Identify and test all interdependent components
5. Develop and conduct strenuous stress and load tests
6. Test integrated solution at different times and in all environments
7. Allow for out of coverage with wireless communications
8. Keep the mobile consumer involved
9. Plan for end-to-end support and maintenance
10. Develop a detailed backup plan
11. Identify and maintain vendor technical support
12. Roll out in stages

Section 3:

Case studies

Case Studies

Wireless computing systems are gaining increasing popularity as companies realize the benefits this technology can deliver. Some systems have been well publicized such as those in use by United Parcel Service, Otis Elevator, Federal Express, General Electric, and Sealand. Applications for wireless computing extend far beyond the popular systems that have been well publicized, however. Many are finding creative ways to use new technologies to enhance their business operations.

The following case studies represent a cross section of technologies and products which have been used to develop and implement successful wireless computing systems. The companies included in these case studies are:

- Viacom
- Avis
- Coffee, Sugar, and Cocoa Exchange
- Spaulding and Slye
- Phoenix Transportation and Taxee, Inc.
- Physicians Sales and Service

Case Study 1: Viacom

INDUSTRY: Field Service

APPLICATION: Field Service Dispatch, Scheduling and Billing

COMPANY: Viacom

NUMBER OF MOBILE CONSUMERS: 89 technicians (12 in phase I)

BUSINESS GOALS and BENEFITS:

- Improves operational output per technician and overall efficiency
- Also improves customer service and sales abilities
- Shorter billing cycle and improved cash flow
- Reduced inventory losses
- Improved control of field operations allows managers and dispatchers to function pro-actively to customer service needs

DESCRIPTION: Viacom uses Ubiquinet's Cable TV Resource Management System (CTRMS) to provide dispatch and scheduling for its field service force while concurrently automating the billing process. This system also integrates with third-party billing vendors like CableData (a subsidiary of U.S. Computer Services. Mapping and inventory management capabilities are also included. Bar coding allows them to track the locations of converters and remotes.

Field technicians receive service assignments through the mobile computer. The dispatching and mapping programs work in tandem with CableData. The system considers several variables including the technician's location, skill level, overtime probabilities. System tracks the technicians' available time and dispatches the next available best-fit job in the queue. Therefore, the best available technician is sent at the least possible cost to Viacom. Once the technician has reached the customer location, they are prepared to begin the service call. The field technicians enter mobile data pressing buttons on the pen-based mobile computer. Field technicians save considerable time by not having to queue up on the radio, or use the customer's phone, to call in service information to the dispatcher. If a converter or remote control unit is required for the service call, the technician scans its identification number using a bar code wand. This information is automatically entered into the customer record which has been sent to the mobile computer. After the job has been completed, the technician secures the customer's verification and signature directly on the computer screen. The technician sends information to Viacom's host computer for processing. Viacom receives immediate notification of what service was performed and which equipment used. Additionally, this information along with the customer's verification is sent to the billing system so that invoices can be sent out without delay. The greatest efficiency gain is in the process of closing jobs.

PILOT DURATION: about 3 months (2 weeks in the future)

ROLLOUT DURATION: about 2 months (3 weeks in the future)

VENDOR: Ubiquinet

SOFTWARE:

Mobile Sites—Proprietary application written in C

Host Site(s)—Keyware, CableData, Oracle version 7

HARDWARE:

Mobile Sites—Telxon

Host Site(s)—Tandem with OS/2 front-end personal computer

NETWORK(S):

Mobile Sites—ARDIS

Host Site(s)—Novell LAN with OS/2 workstations

Case Study 2: Avis

INDUSTRY: Automobile Rental

APPLICATION: Customer Check-In

COMPANY: Avis

NUMBER OF MOBILE CONSUMERS: 175+ (more than 1 driver per bus)

BUSINESS GOALS and BENEFITS:

- Increased operational efficiency
- Maintaining a high level of customer satisfaction
- Increased job satisfaction

DESCRIPTION: The Customer Check-in wireless computing system automates the process of transmitting information from Avis' Wizard mainframe at its

World Headquarters to Wizard on Wheels (WOW) mobile data terminals which are located on the airport shuttle buses. This system provides comprehensive rental information for tens of thousands of Preferred Customers at 16 major airport locations operating more than 175 shuttle buses.

Shuttle bus drivers enter the customer identification on the Wizard on Wheels (WOW) terminal. Using the ARDIS network, information is sent to the Wizard mainframe. Within seconds, the mainframe retrieves the customer's reservation information, selects the appropriate vehicle from the location's automated ready-line database, and prints a rental agreement at the Avis express facility. This facility is co-located with the rental counter. The printed rental agreement is placed in the appropriate vehicle by an Avis employee. The bus driver is simultaneously sent information indicating what space the customer's car is parked in as well as the make, model, and color of the car. The shuttle bus driver then delivers the customer to that car. The rental agreement is usually ready by the time the customer reaches the car.

The estimated project cost was more than $1 million for implementation, plus an additional $100,000 per year to support the system.

PILOT DURATION: N/A

ROLLOUT DURATION: N/A

SOFTWARE:

Mobile Sites—N/A

Host Site(s)—N/A

HARDWARE:

Mobile Sites—Motorola MDT 9100-11

Host Site(s)—N/A

NETWORK(S):

Mobile Sites—ARDIS

Host Site(s)—ARDIS and N/A

Case Study 3: Coffee, Sugar, and Cocoa Exchange

INDUSTRY: Financial Services

APPLICATION: Price Reporting

COMPANY: Coffee, Sugar & Cocoa Exchange

NUMBER OF MOBILE CONSUMERS: 36

BUSINESS GOALS and BENEFITS: The Automated Price Reporting System allowed the Coffee, Sugar, and Cocoa Exchange to streamline the process and realize several benefits including:

- Reduced errors and associated cost savings
- Increased speed of data transfer
- Reduced headcount and associated Expense
- Increased SEC compliance
- Production of an audit trail

DESCRIPTION: The Price Reporting System automates the process of reporting a trade. Prior to implementation of the system, a trade was reported in the following way:

1. A Price Reporter receives a signaled price from a trader.
2. The Price Reporter writes the commodity, quantity, and price on a piece of paper which is then handed to a runner.
3. The runner delivers the piece of paper to a catcher.
4. The catcher delivers it to a data entry person who types the information into a terminal.
5. The terminal is connected to a Tandem host, which records the trade and posts it on the board.

The Price Reporting System places a handheld computing device with wireless communications in the hands of the Price Reporter, enabling them to immediately deliver the information received from the trader to the host computer. The wireless communication link performs the runner and catcher functions. The trade is captured in an electronic format, thus also eliminating the need for the terminal data entry. As a result of the elimination of the runner, catcher, and data entry functions, headcount and its associated expense is saved. From the time the price reporter hits the send button, to the time the data is received by the Tandem, no more than 2 seconds have elapsed. This system provides connectivity to the Tandem host as well as 100% redundancy between the handheld computer and the Tandem host.

PILOT DURATION: 3 months

ROLLOUT DURATION: 45 days

SOFTWARE:

Mobile Sites —Proprietary Written in "C"

Host Site(s) —Written in Tandem COBOL 85 with Tandem Pahway subsytem and Tandem SCOBOL (Screen COBOL)

HARDWARE:

Mobile Sites—Norand

Host Site(s)—Tandem

NETWORK(S):

Mobile Sites—Norand Spread Spectrum Local Area Network

Host Site(s)—Tandem 3270 emulator connected to a SNA line connected to a Norand controller connected to the Norand local area network

Case Study 4: Spaulding and Slye

INDUSTRY: Real Estate Services

APPLICATION: Construction Project Management

COMPANY: Spaulding and Slye

NUMBER OF MOBILE CONSUMERS: 9

BUSINESS GOALS and BENEFITS:

- Offers competitive differentiation
- Allows a company to keep on top of the business
- Saves time and money
- Clients can become an integral part of the project

DESCRIPTION: Spaulding and Slye uses Lotus Notes, Lotus workgroup applications, and SkyTel's wireless messaging services to manage construction projects. This 29-year-old Fortune 100 firm manages major construction projects, helps corporations develop real estate strategies, leases and sells commercial space, and provides property and asset management services.

Typical development sites are for buildings that run upwards of 100,000 square feet, cost tens of millions of dollars, and are located in cities as far away as San Juan, Puerto Rico and San Diego, California. Traditional project management methods include lengthy telephone calls, long typewritten reports, and multiple person-to-person meetings. Lotus Notes allows them to integrate all aspects of project management into a single information and communications system. Project team members are often scattered between huge construction sites and sometimes work several stories up in steel skeletal structures while others are in offices in remote locations. SkyTel pagers are the mobile devices which allow project team members to receive messages and time-critical information, including updates when milestones were reached or missed. The SkyTel pager is accessed via the Lotus Notes SkyTel Pager Gateway and the SkyTel wireless communications network. Information in sent from the Lotus Notes database to project team members in remote locations. Architects can send plans, engineers can track supply deliveries, change orders can be circulated, and clients can check on the status of the project. Databases monitor key project events and automatically transmit messages via the SkyTel network. When a critical event takes place, the system composes a message, dials up the SkyTel network, and sends the message for transmission to all team members using SkyTel pagers. Messages are received within a minute of transmission. All project members may compose and send a text message from within Notes to any SkyTel pager user.

PILOT DURATION: No pilot—12 month development project

ROLLOUT DURATION: 2 weeks

SOFTWARE:

Mobile Sites—None required

Host Site(s)—Lotus Notes and workgroup applications

HARDWARE:

Mobile Sites—SkyTel pager

Host Site(s)—IBM 486 personal computer

NETWORK(S):

Mobile Sites—SkyTel wireless communications network

Host Site(s)—Banyan local area network

Case Study 5: Phoenix Transportation and Taxee, Inc.

INDUSTRY: Consumer Service

APPLICATION: Credit Card Verification

COMPANY: Phoenix Transportation and Taxee, Inc.

NUMBER OF MOBILE CONSUMERS: 7

BUSINESS GOALS & BENEFITS:

- Substantial (about 20% in 3 months) Increase in Revenue
- Fast credit verification
- Flexibility in the location of POS terminals

DESCRIPTION: Taxi drivers use Firstnet Corporation's AireTrans Transaction Air Connect System to verify customers' credit cards. AireTrans provides a cost-effective alternative for businesses (e.g., taxis, delivery vehicles) who could not previously offer customers the convenience of paying by credit card. Prior to using this service, drivers could not service customers unless they paid cash. The drivers can now wirelessly send and receive information about credit limits or approvals in less than six seconds. Although landline connections are not an option, they may take up to 20 seconds. The airlink is encrypted for data security. AireTrans transaction fees are comparable to those using standard landline.

PILOT DURATION: 6 months

ROLLOUT DURATION: 1 month

SOFTWARE:

Mobile Sites—Aire Trans

Host Site(s)—Firstnet Corporation's propietary

HARDWARE:

Mobile Sites—VeriFone Trans 460 point of sale terminal Cincinnati Microwave Inc. MC-DART CDPD modem

Host Site(s)—GenSar Technologies Inc.'s Tandem Computer

NETWORK(S):

Mobile Sites—Bell Atlantic Mobile AirBridge℠ Packet service

Host Site(s)—Gensar Technologies Inc.'s network from Transaction Network System (TNS)

Case Study 6: Physicians Sales & Service

INDUSTRY: Health Care

APPLICATION: Salesforce Automation

COMPANY: Physicians Sales & Service

NUMBER OF MOBILE CONSUMERS: 412

BUSINESS GOALS & BENEFITS:

- Same day service for customers
- Increased Sales Volume 15% in 18 months
- Shorter Delivery Cycle
- Lower Administrative Headcount

DESCRIPTION: As a member of the sales team that fuels an annual growth rate of over 40%, PSS salespeople handle 200 to 300 accounts each and are expected to make about 30 sales calls a day. To assist them in their efforts, PSS equips sales representatives with a salesforce automation system which incorporates wireless computing technology. When a salesrep enters a physician's office they can call up the doctor's purchase history, scour the company's inventory and order from any warehouse in its system. Salesreps can also give demonstrations using the notebook computer. PSS is acknowledged as the first medical supply company that has introduced a notebook computer for customer information and order entry.

Sales representatives use customized software, notebook computers, radio modems and the RAM Mobile Data wireless network to link directly to the office server. In minutes the sales representatative can verify product availability, delivery dates and confirm the order. Orders are transmitted to the office and broadcasted back to the field for approval prior to billing. Information provided by the salesperson helps to facilitate and speed up the fulfillment process. PSS can now use information from the field salesperson to post the order, invoice the customer and ship the order. PSS has reduced the number of administrative personnel that were part of the fulfillment process while decreasing the delivery cycle for its orders.

A major benefit that PSS has realized since implementation of the system is that it can provide same day service for its customers. This increased level of service coupled with greater productivity of its salesforce has helped it achieve a 15% increase in revenue within 18 months.

PILOT DURATION: 6 weeks

ROLLOUT DURATION: 12 months

SOFTWARE:

Mobile Sites—Proprietary from Thinque Corporation

Host Site(s)—Proprietary Alpha Micro minicomputer

HARDWARE:

Mobile Sites—COMPAQ Concerto

Host Site(s)—COMPAQ Prolinea

NETWORK(S)

Mobile Sites—RAM Mobile Data

Host Sites(s)—WINDOWS NT

Index

A

B

C

D

G

H

I

J

K

L

M

N

O

R

S

T

U

V

W

X

POSTAGE
REQUIRED

ACT Inc.
P.O. Box 978
South Orange, NJ 07079